THE SERIES OF
"CHINA'S MARITIME DEVELOPMENT"

THE MARITIME UNDERTAKING OF CHINA

AUTHORS: JIA YU, ZHANG DAN

China Intercontinental Press

图书在版编目（CIP）数据
中国的大洋事业：英文 / 张丹，贾宇编著；李斯雅等译. -- 北京：五洲传播出版社，2014.9（中国海洋丛书）
ISBN 978-7-5085-2944-8

Ⅰ. ①中… Ⅱ. ①张… ②贾… ③李… Ⅲ. ①海洋经济–经济发展–研究–中国–英文 Ⅳ. ① P74

中国版本图书馆 CIP 数据核字 (2014) 第 256801 号

“中国海洋”丛书

策　　划：付　平
出 版 人：荆孝敏
主　　编：张海文 高之国 贾　宇

中国的大洋事业（英文）

编　　著：张　丹 贾　宇
责任编辑：黄金敏 姜　超
翻　　译：魏　怡 李斯雅
图片提供：国家海洋局海洋发展战略研究所 东方 IC CFP
装帧设计：丰饶文化传播有限责任公司
出版发行：五洲传播出版社
社　　址：北京市北三环中路 31 号生产力大楼 B 座 7 层
电　　话：0086-10-82007837（发行部）
邮　　编：100088
网　　址：http://www.cicc.org.cn　http://www.thatsbooks.com
印　　刷：北京市艺辉印刷有限公司
开　　本：889mm×1194mm　1/32
字　　数：65 千字
图　　数：80 幅
印　　张：3.75
印　　数：1—5000
版　　次：2014 年 12 月第 1 版第 1 次印刷
定　　价：72.00 元

Foreword

"Clasping the moon in the ninth heaven while seizing turtles deep down in the five oceans" is the long-cherished dream of generations of the Chinese people. With the increasing comprehensive national power (CNP) and the implementation of maritime power strategy, China's dream of going out to the oceans is no longer faraway. *The Maritime Undertaking of China* is one of the series of *China's Maritime Development* planned and published by China Intercontinental Press. This book introduces the key concepts, development history and the maritime undertaking that China is embarking upon, such as the scientific expedition, participation in international seabed affairs and application of international seabed mining areas. This book is compiled by the Marine Development Strategy Institute of State Oceanic Administration. The author is Zhang Dan and the editor for the final draft is Jia Yu.

Contents

KNOWING INTERNATIONAL SEABED AREA

Enormous Treasures

In the *United Nations Convention on the Law of the Sea* (UNCLOS), the seabed and ocean floor and subsoil beyond the national jurisdiction is called the Area. As the constitution of oceans, the *United Nations Convention on the Law of the Sea* (UNCLOS) adopted in 1982 established the legal order of international oceans and played a significant role in the peaceful use of the sea and marine environment protection.

UNCLOS divides international oceans into various areas of different legal status and jurisdiction system: (1) sea areas under national jurisdiction: internal waters (inland sea), territorial waters, contiguous zone, Exclusive Economic Zone (EEZ) and continental shelf; (2) areas beyond national jurisdiction (ABNJ): high seas, the Area and other sea areas beyond the jurisdiction of the coastal State.

High seas cover all parts of the sea that are not included in the exclusive economic zone, in the territorial sea or in the internal waters of a State, or in the archipelagic waters of an archipelagic State. The Area refers to the seabed and ocean floor and subsoil beyond the limits of national jurisdiction. Both beyond national jurisdiction, the limits of high seas and the Area are not in complete due to the system of the continental shelf beyond 200 nautical miles from the baseline from which the breadth of the territorial sea is measured.

According to UNCLOS, the continental shelf of a coastal

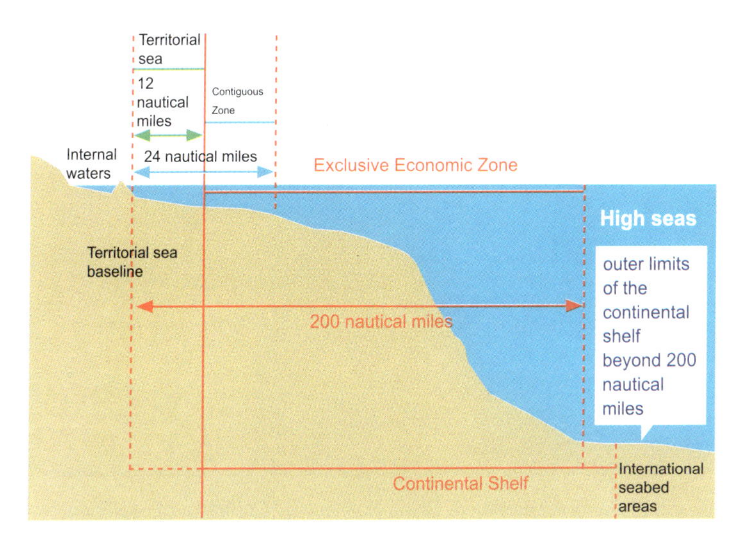

Various Sea Areas with in the Framework of UNCLOS

State comprises the seabed and subsoil of the submarine areas that extend beyond its territorial sea throughout the natural prolongation of its land territory to the outer edge of the continental margin does not extend up to that distance. The continental margin, or to a distance of 200 nautical miles from the baselines from which the breadth of the territorial sea is measured where the outer edge of the continental margin does not extend up to that distance. When the continental shelf extends over 200 nautical miles, the outer limits of continental shelf can extend beyond 350 nautical miles and even farther. This is often abbreviated as the "outer continental shelf". Due to the system of outer continental shelf, the limits of high seas and those of the Area differ to some extent. The superjacent waters of the Area are undoubtedly high seas. The seabed and ocean floor and the subsoil of the submarine area of high seas could either be within the limits of the Area or those of

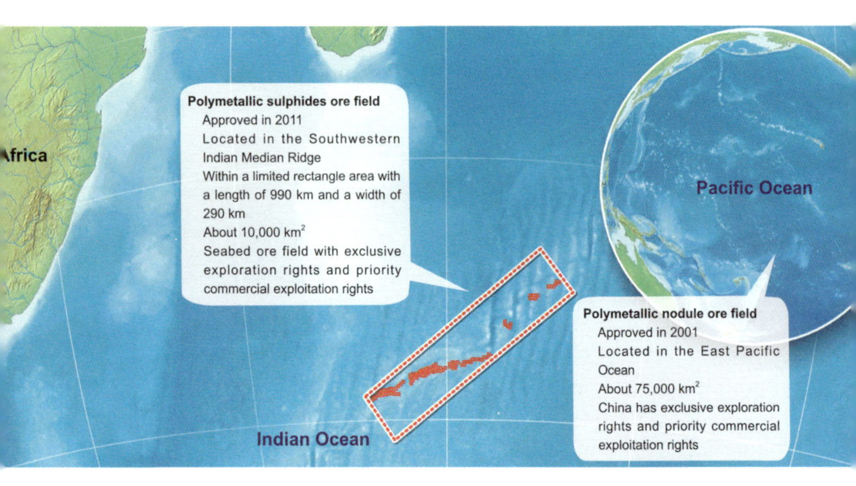

Seabed Mineral Areas Where China Has Gained Contracts of Exploring

the outer continental shelf of the coastal State. According to the statistics of some scholars, the Area occupies about 65% of the world's oceans. [1]

According to UNCLOS, information on the limits of the continental shelf beyond 200 nautical miles from the baselines from which the breadth of the territorial sea is measured shall be submitted by the coastal State to the Commission on the Limits of the Continental Shelf set up under Annex II on the basis of equitable geographical representation. The Commission shall make recommendations to coastal States on matters related to the establishment of the outer limits of their continental shelf. The limits of the shelf established by a coastal State on the basis of these recommendations shall be final and binding.

---

1. Wei Min, ed., *The Law of the Sea*, 1st edition, Beijing: Law Press China, 1987, p 218.

It is highly complicated to decide technically the outer limits of continental shelf for the coastal State. They should not only have corresponding science and technology, but also carry out a great number of on-the-spot investigations and data collecting. The consideration of the data and other materials submitted by coastal States concerning the outer limits of the continental shelf by the Commission on the Limits of the Continental Shelf is no easy matter. A huge amount of materials are involved. Thus, this work is rather time-consuming. Since closely connected with the limits of the continental shelf of coastal States, the limits of the Area cannot be settled in a short period of time.

The Area abounds in rich resources.[1] Polymetallic nodules is one of the main resources that human beings have discovered in the Area and also one of the earliest discovered resources of the Area. The emergence of the legal status of the Area is also directly related with the large-scale discovery of polymetallic nodules resources. This is also why UNCLOS highlights in Article 133 in particular "including polymetallic nodules" when defining the resources of the Area.

The increasing human activities contribute to the discovery of various resources in the Area, mainly including polymetallic sulphides, cobalt-rich crusts and natural gas hydrates. Besides mineral resources, diverse species and biological communities were discovered there, which have provided invaluable deep sea biological genetic resources and brought significant commercial values.

1. Article 133, UNCLOS

Polymetallic Nodules: Important Strategic Resources in Deep-sea Areas

Polymetallic Nodules

Polymetallic nodules were first discovered in the Kara Sea in the Arctic Ocean off Siberia in 1868. During the scientific expeditions of the H.M.S. Challenger (1872-76), they were found to occur in most oceans of the world.

Polymetallic nodules, also called manganese nodules, are rock concretions formed of concentric layers of iron and manganese hydroxide around a core. The core may be microscopically small and is sometimes completely transformed into manganese mineral by crystallization. When visible to the naked eye, it can be a small test (shell) of microfossil (radiolarian or foraminifer), a phosphatized tooth of shark, basalt debris or even fragments of earlier nodules. The thickness and regularity of the concentric layers are determined by the successive stages of growth. On some nodules they are discontinuous, with noticeable differences between the two sides.

Nodules vary in size from tiny particles visible only under a microscope to large pellets more than 20 centimeters across. However, most nodules are between 5 and 10 cm in diameter, about the size of potatoes. Their surface is generally smooth, sometimes rough, mammilated (knobby) or otherwise irregular.

The bottom, buried in sediment, is generally rougher than the top.[1]

Nodules have been found in all the oceans and even in lakes. However, nodules of economic interest are more localized. Three areas have been selected by industrial explorers: the Clarion-Clipperton Fracture Zone (CCFZ) in the Northeast Pacific Ocean, the Peru Basin in the Southeast Pacific Ocean and the central area of the North Indian Ocean. Polymetallic nodules are now known to contain six kinds of mineral materials: manganese nodules, iron concretion, siliceous nodules and carbon nodules, etc. The total amount of polymetallic nodules lying on the sea floor is estimated about 500 billion tons.[2]

The chemical composition of nodules varies according to the kind of manganese minerals and the size and characteristics of the core. Those of economic interest have the following constituents: manganese (29%), iron (6%), silicon (5%), and aluminum (3%). Nickel (1.4%), copper (1.3%) and cobalt (0.25%) are the most valuable.

Manganese, copper, nickel, cobalt and other mineral substances are all significant strategic resources. Manganese is widely used in iron and steel industry, oil and natural gas drilling and transportation. Nickel is a very scarce mineral resource and the main substance of making superalloy. Copper is widely employed in the manufacture of military and electrical equipment. Cobalt is a principal substance of superalloy used in the manufacture of the engine of jet planes and combustion gas turbine. Without these mineral substances, modern top-end industries and state-of-the-

1. http://china-isa.jm.chineseembassy.org/chn/gjhd/hdzy/t218968.htm
2. Research Group for *China's Ocean Development Report*, *China's Ocean Development Report 2012*, Beijing: China Ocean Press, P198.

art scientific research would be an empty talk. Presently, there are different estimates about the mineral reserves of polymetallic nodules under commercial exploration under current conditions ranging from 480 million to 13.5 billion tons.[1]

Polymetallic Sulphides: Hydrothermal Liquid Spouted out of "Black Smoker"

In 1979, on the East Pacific Rise (EPR) at 21 degrees north latitude off Baja California (Mexico), scientists exploring the ocean floor discovered chimney-like formations of dark rock atop sulphide mounds, spewing hydrothermal liquid and surrounded by animal species different from any previously known. Since then, studies have shown that these black-smoker complexes are an outgrowth of the formation of new oceanic crust through seafloor spreading as the tectonic plates underlying the earth's surface converge or move apart. Moreover, this activity is intimately associated with the generation of metallic mineral deposits at the seafloor.[2]

At water depths up to 3,700 meters, hydrothermal fluids, having seeped from the ocean into subterranean chambers where they are heated by the molten rock (magma) beneath the crust, are discharged from the black smokers at temperatures up to 400° Celsius. As these fluids mix with the cold surrounding seawater, metal sulphides in the water are precipitated onto the chimneys and nearby seabed.

Most polymetallic sulphide deposits are found at sites in the

1. Research Group for *China's Ocean Development Report*, *China's Ocean Development Report 2012,*Beijing: China Ocean Press, P198-199.
2. http://china-isa.jm.chineseembassy.org/chn/gjhd/hdzy/t218969.htm

Black Smoker

mid-ocean areas, distributed in the East Pacific Rise, the Southeast Pacific Rise and the Northeast Pacific Rise. Several deposits are also known to distribute on the Mid-Atlantic Ridge and the submarine ridge of the Indian Ocean. There are ocean ridges of 60,000 kilometers in total worldwide, but only a few of them has been explored and investigated.

High concentrations of base metals (copper, zinc, lead) and precious metals (gold, silver) in some of these polymetallic sulphide deposits have recently attracted the interest of the international mining industry. The scales of the mine beds of polymetallic sulphides are generally quite small with a few exceptions. For example, in a mine bed of polymetallic sulphides in Galapagos Ridge (height: 1000 meters, width: 200 meters, length: 35meters) the average contents of iron, copper and zinc are 35%, 10% and 0.1% respectively. Silver, cadmium, vanadium and tin are also contained. The estimated reserve is about 25 million tons, with a total value of 3.9 billion dollars.[1] Compared to polymetallic nodules and cobalt-rich crusts, the mine beds of polymetallic sulphides are characterized by shallow water depth of deposits (from several hundreds of meters to 2,000 meters), high grades of concentration of orebody, and low difficulties in exploring and smelting, and other merits.

Cobalt-rich Crusts: Concentrating in the Seamounts

Oxidized deposits of cobalt-rich ferromanganese crusts are found throughout the global oceans on the flanks and summits of seamounts, ridges and plateaus. There are about 50,000 seamounts in the Pacific, where the richest deposits of cobalt-rich crusts are found. The first systematic investigation of crusts was carried out in 1981 in the mid-Pacific Ocean. Early scientific research was conducted by scientific research teams from Germany, the United States, the Union of Soviet Socialist Republics (later the Russian Federation), Japan, France, the United Kingdom, China and the South Korea.

The minerals in cobalt-rich crusts have precipitated out of the cold ambient seawater and onto the rock surface, likely with the aid of bacterial activity. Crusts do not form in areas where sediment covers the rock surface. The crusts form pavements up to 25 centimeters thick and cover an area of many square kilometers. They are found at water depths of about 400-4,000 meters.

According to one estimate, about 6.35 million square

1. Fouquet Y., U.Von Stackelberg, J.L.Charlou, P.M.Herzig, R.Muhe and M.Wiedicke, 1993, *Metallogenesis in Back-arc Environments: the Lau Basin Example. Economic Geology*, 88, 2154-2181.

Cobalt-rich Crusts

kilometers, or 1.7 percent of the ocean floor, is covered by cobalt-rich crusts, translating to some 1 billion tons of cobalt. Based on grade, tonnage and oceanographic conditions, the central equatorial Pacific region offers the best potential for crust mining, particularly the exclusive economic zones around Johnston Island and Hawaii of the United States, the Marshall Islands, the Federated States of Micronesia and the Area of the mid-Pacific.

Crusts contain a high content of cobalt, up to 1.7 percent, and large areas of individual seamounts may contain crusts with average cobalt content of up to 1%. These cobalt proportions are much higher than in land-based ores, which range from 0.1 to 0.2 % cobalt. Other than cobalt, the most valuable of the crust metals are titanium, cerium, nickel and zirconium, in that order.

The types of metals occurring in cobalt-rich crusts – notably cobalt, manganese and nickel – are used to add specific properties to steel, such as hardness, strength and resistance to corrosion.

In industrial countries, between one fourth and one half of cobalt consumption is used by the aerospace industry in superalloys. These metals are also employed in chemical and high-technology industries, for such products as photovoltaic and solar cells, superconductors, advanced laser systems, catalysts, fuel cells and powerful magnets, as well as for cutting tools.[1] Comparing to polymetallic nodules and polymetallic sulphide, cobalt-rich crusts can be explored easier with similar smelting process.

Besides, deep seabed areas abound in rich oil and natural gas with a known reserve of 250 billion tons, which occupies over one third of the total amount of the world's oil and natural gas. Among these, a considerable proportion is within the Area.

Distribution of Metallic Mineral Resources in the Area

	Polymetallic Nodules	**Cobalt-rich Crusts**	**Polymetallic Sulphides**
Distribution Area	Ocean Basin	Seamount	Ocean Ridge, Back-Arc Basin
Depth of Water	4000-6000 meters	800-2500 meters	500-2000 meters
Potential Resources	~70 billion tons	~21billion tons	~0.4 billion tons
Useful Components	Copper, Nickel, Cobalt, Manganese	Cobalt, Nickel, Lead, Platinum	Copper, Lead, Zinc, Gold, Silver
Deposit Form	Two-Dimensions	-	Three-Dimensions

1. http://china-isa.jm.chineseembassy.org/chn/gjhd/hdzy/t218970.htm

Biological Genetic Resources: New Understanding of "Lifeless Desert"

People used to think that deep sea is a lifeless desert since living things are hard to survive. Now, human beings have realized that there are very rich species in deep seabed areas. Deep-sea biological genetic resources have aroused international interest as a new kind of resource. Biological communities are largely spread around hydrothermal vents along the ocean ridge, including tubeworms, mussels, clams and shrimps, along with other widespread micro-orgasms. Widely distributed seamounts are like islands for biological communities. Communities of hairy annelids, nematodes and foraminifers are distributed in the sediment of vast ocean basins.

Living in unique physical, chemical and biological environment and especially in high-pressure and extremely changing temperature gradients and surrounded by toxic substances, deep-sea organisms develop peculiar biogenetic textures and metabolism. Special biological active substances have formed inside these organisms such as various extremozymes where are basophil, pressure-resistant, thermophilic, cryophilic and antitoxic. The diverse functions of these biological active substances are the deep-sea biological resources of the most application value.

It is predicted that deep-sea biological resources will be extensively used in fields like industry, medicine and environmental protection. Till now, the application of deep-sea biological genetic resources in the world has generated industrial values of billions of dollars.

From the perspective of biodiversity and ecosystem, there

are diverse biological communities and genes in deep-sea areas. As a highly sensitive, very scarce and vulnerable ecosystem, the hydrothermal-vents and some parts of seamounts have unique scientific research values which require special attention and protection of the whole international community. [1]

Common Heritage of Mankind

The Area is a brand new concept of international law established by UNCLOS. The legal status of the Area is neither similar to areas under national jurisdiction like territorial waters, EEZs, continental shelf and others, nor to high seas where states have limited rights on.

Although both beyond the national jurisdiction, high seas and the Area are different from each other with regard to their legal status. High seas are open to all States, whether coastal or land-locked. Freedom of the high seas comprises both for coastal and land-locked States: freedom of navigation, freedom of overflight, freedom to lay submarine cables and pipelines, freedom to construct artificial islands and other installations permitted under international law, freedom of fishing and freedom of scientific research.

In UNCLOS, there are special provisions about the legal system of the Area. The basic principles are as follows. The Area and its resources are the common heritage of mankind. No State

1. http://china-isa.jm.chineseembassy.org/chn/gjhd/hdzy/t218972.htm

Seabed World

shall claim or exercise sovereignty or sovereign rights over any part of the Area or its resources, nor shall any State or natural or juridical person appropriate any part thereof. No such claim or exercise of sovereignty or sovereign rights nor such appropriation shall be recognized. All rights in the resources of the Area are vested in mankind as a whole, on whose behalf the Authority shall act. The Authority shall provide for the equitable sharing of financial and other economic benefits derived from the activities of exploration and exploitation in the Area through any appropriate mechanism for all States.

International Seabed Authority

International Seabed Authority (ISA) is an international organization established by UNCLOS to administer the activities in the Area. It was formally established in 1994 headquartered in Kingston, Jamaica. It has four main organs including the Assembly, the Council, the Secretariat and the Enterprise.

The Assembly is the organ of supreme power of the Authority. It consists of all the members of the Authority, which meets in regular annual sessions and in such special sessions as may be decided by the Assembly, or convened by the Secretary-General at the request of the Council or of a majority of the members of the Authority. The Council consists of 36 members of the Authority elected by the Assembly, including four members from among those States Parties which have either consumed more than 2 per cent of total world consumption or have had net imports of more than 2 per cent of total world imports of the commodities produced from the categories of minerals to be derived from the Area, as well as the largest consumer (a); four members from among the eight States Parties which have the largest investments in preparation for and in the conduct of activities in the Area (b); four members from among States Parties which on the basis of production in areas under their jurisdiction are major net exporters of the categories of minerals to be derived from the Area, including at least two developing States whose exports of such minerals have a substantial bearing upon their economies (c); six members from

ISA Headquarters

among developing States Parties, representing special interests (d); eighteen members elected according to the principle of ensuring an equitable geographical distribution of seats in the Council as a whole (e). The Secretariat of the Authority performs administrative functions of the Assembly, the Council and any subsidiary organ. The Enterprise is an organ of the Authority, empowered to conduct seabed mining. The Secretariat is now operating on the behalf of the Enterprise. A temporary Director-General is appointed by the Secretary-General of the Authority to monitor the Secretariat in performing this function until the Enterprise is functional and independent from the Secretariat.

Besides, ISA has two subsidiary organs: the Legal and Technical Commission (LTC) and the Economic Planning Commission. Until May 25th 2014, 166 members are included in the Authority.

INTERPRETING THE LEGAL SYSTEM OF INTERNATIONAL SEABED AREA

Disputes over the Legal Status of the Area

The ocean was traditionally divided into two parts: territorial waters and high seas. Nowadays, with the enhancement people's understanding, development and utilization of oceans it has become a trend for coastal States to enlarge their jurisdiction over oceans. This trend has been more remarkable especially since WWII.

Coastal States' struggle over maritime jurisdiction has extended from water surface to deep sea. International sea-bed

the Laying of China's Largest Cross-section Submarine Cables in the Sea Area Around Xiamen

Hugo Grotius,
Father of the International Law

area originally belonging to high seas has also become a target of struggle. Against such background, the legal status of international seabed area has emerged and become more prominent. Besides the "common heritage of mankind" principle finally determined by UNCLOS, three principles were the most representative ones in early discussions about the legal status of the Area, including: res nullius theory, res publica theory (public property theory) and theory of the freedom of the high seas.

According to Roman law, res nullius refers to an object that can be owned and is not yet the object of rights of any specific object. Such ownerless property are usually free to be owned and can become someone's private property by means of possession. Terra nullius, an application of res nullius, can also become someone's private property by means of occupation.

Res publica (public property) is something held in common by all people and cannot be used or enjoyed by anyone.

Hugo Grotius, the renowned jurist of the Dutch Republic and father of international law published the famous *Mare Liberum*, a Latin title that translates to *The Free Sea* in 1609. He held that sea can neither be picked up nor enclosed nor exhausted after using up. Therefore, it is international territory belonging to no nation.

In 1958, the *Convention on the High Seas* was signed at the

First United Nations Conference on the Law of the Sea in which the principle of the freedom of the high seas was established.

Opinions mentioned above about the legal status of the Area and the resources in it were not adopted by UNCLOS. But with the joint efforts of developing nations, the principle of "common heritage of all mankind" was finally listed in UNCLOS, making it the fundamental legal principle regulating the special political and geographical unit, the Area.

In August, 1967, Arvid Pardo, Malta's Ambassador to the United Nations, requested the inclusion of the following new item concerning the development of the seabed in the agenda at the 22nd Session of the General Assembly: "Declaration and treaty concerning the reservation exclusively for peaceful purpose of the sea-bed and of the ocean floor, underlying the seas beyond the limits of present national jurisdiction, and the use of their resources in the interests of mankind", with an explanatory memorandum attached. This is the first time when "common heritage of mankind" was mentioned officially.

Pardo considered that the time had come to declare the ocean floor to be the common heritage of mankind and proposed that immediate steps should be taken to draft a treaty. Eventually, the 22nd Session of the General Assembly adopted Resolution 2340 "decided to establish an Ad Hoc Committee to study the peaceful uses of the seabed and the ocean floor beyond the limits of national jurisdiction…to study the scope and various aspects of this item".

Developing countries held that it was necessary to build a regime of the Area different from that of high seas and conduct international control over the prospecting, exploration and exploitation in the Area for the benefit of all mankind. This proposal is opposed to the theory of a few maritime powers to plunder the resources in the Area. Some countries from plundering resources of the Area.

In 1969, the 24th Session of the General Assembly adopted Resolution 2574. It declared that pending the establishment of an international regime concerning the Area, "states and persons, physical or juridical, are bound to refrain from all activities of exploitation of the resources of the area of the sea-bed and ocean floor and the subsoil thereof, beyond the limits of national jurisdiction; No claim to any part of that area or its resources shall be recognized."

In 1970, the 25th Session of the General Assembly adopted Resolution 2749: *Declaration of Principles Governing the Seabed and the Ocean Floor, and the Subsoil Thereof, beyond the Limits of National Jurisdiction*, which made a great difference in the formation and development of the legal regime in the Area. It introduced the legal status of the Area and the legal principles of seabed activities "the seabed and ocean floor, and the subsoil thereof, beyond the limits of national jurisdiction (hereinafter referred to as the area), as well as the resources of the area, are the common heritage of mankind."

The passing of the *Declaration* suggested that the "common heritage of mankind" principle had been accepted by most countries of the world. Finally, the "common heritage of mankind" principle was listed in Article 136 in UNCLOS, assuring that the Area is the "common heritage of mankind". This was the first time that this concept was formulated in the form of an article. To commemorate his contribution to the birth of modern law of the sea, Malta's Ambassador Arvid Pardo was named "Father of

Father of the Law of the Sea Conference: Arvid Pardo

the Conference of the Law of the Sea".

The "common heritage of mankind" mostly includes the following contents.

(1) Owned by All Mankind. The Area and its resources belong to all mankind and the entire international community rather than any state, natural person or juridical person. Instead of referring to any single nation, natural person or legal person, the mankind here includes both the citizens of all nations today and their offspring in the future. All of them have the right to participate in the activities in the Area and share the benefits and profits from the activities.

(2) Managed and Used by All Mankind. "Used by mankind" includes joint management and joint exploitation. The Area and its resources are owned by all human beings, thus, all states should manage and exploit them together. All States have agreed on establishing a new authority operating under the authorization of all members to control the activities in the Area. Activities in the Area shall be organized, carried out and controlled by the Authority on behalf of mankind as a whole. ISA was set up accordingly to exercise rights on the Area on behalf of all mankind. Besides, UNCLOS also founded the Enterprise to directly participate in prospecting, exploring and exploiting the resources in the Area.

(3) Shared by All Mankind. The Area and its resources are

common heritage of mankind. Therefore, the exploration and exploitation of those resources should take into consideration of the benefits of all states. UNCLOS stipulated in express terms that financial interests acquired from activities in the Area, along with some other economic interests, should be distributed on the basis of objective and non-discriminatory standards. Yet a fair distribution indicates enjoying benefits gained from the Area rationally rather than sharing interests equally.

During the negotiation process of UNCLOS, there were some major disagreements on the specific prospection and development of the resources in the Area, although all states had reached a consensus quite earlier on the principle of "common heritage of mankind". The assertions put forward were mainly divided into three types, namely single exploitation system, International Registration and Licensing System and Parallel System.

Developing countries highlighted that the resources in the Area is "common heritage of mankind", thus only the authority on the behalf of all human beings could have the right to explore and exploit resources in the Area.

Countries with advanced marine industry like the United States and the former Soviet Union advocated international registration system and proposed to set a loosely organized authority administering the Area. States which have signed contracts with the authority shall have the right to explore and exploit the resources in the Area.

Developed countries stressed that all states had the right to explore and exploit resources of the Area. The authority only functioned as a place to register. As long as the state informed the authority about his activities in the Area, it would feel free to use the resources in the Area. This proposal only allowed the authority few rights with which only negative functions (the size of application area, the expiration date of the application, the fee for registration and some rules related to environmental protection and safety) could be operated.

Besides, some countries stood for an international licensing system. They argued that authority of the Area must strictly accord with the pre-established and detailed international regulations and grant licenses. It cannot make any random decisions.

To break the stalemate, both developing countries and developed countries made some concessions with regard to the exploitation system of the resources in the Area at the third session held in 1975. Developing countries began to accept the proposal that other entities besides the Enterprise could exploit the resources in the Area after getting permission from the authority. This has modified to some extent the single exploration system that the developing countries had always advocated concerning the

Haiyang Shiyou 720 is carrying out 3D oil seismic exploration in South China Sea.

Henry Alfred Kissinger, the former United States Secretary of State

resource exploration system in the Area. This is an embryonic form of the parallel system.

Based on this, the then United States Secretary of State Henry Kissinger proposed a system the parallel system holding that the Enterprise, and both state-owned and private-owned enterprises of contracting states can exploit the deposits in the Area. The UNCLOS finally adopted the parallel system, determining two entities to exploit the resources in the Area: one is the Enterprise of the Authority; the other includes contracting parties or state-owned enterprises, and natural person or legal person possessing the nationality of contracting states and under the guarantee of the contracting states or natural person or legal person effectively controlled by the citizens of the contracting states.

Compromise Despite Disagreements

After the "Area" system established in the UNCLOS was basically finalized, the U.S. requested a new round of review of the draft of UNCLOS at the tenth session of the Conference on the Law of the Sea, proposing in particular the so-called eight items of concerns and standards of measurement concerning the

Area. It questioned about the ISA and its Enterprise, technology transfer system, restrictions on seabed production, the composition the Council of ISA, the censorship of the UNCLOS, financial conditions of contracts and the protection of preliminary investment and so on.

At the request of the U.S. delegation, a vote was initiated on April 30th 1982 to decide whether UNCLOS should be approved. The UNCLOS was passed by an overwhelming majority of 130 vs. 4 with 17 abstentions.

The road to the establishment and implementation of any complete international regime is always filled with thorns, so is the principle of "common property of mankind". UNCLOS was open for signature December 10th 1982. By the signature-closing day of December 9th on 1984, 159 states and entities had signed UNCLOS. Some Western countries like the United States, the United Kingdom and Germany refused to sign it since they were dissatisfied with some articles concerning the Area. According to some resolutions passed at the 3rd United Nations Conference on the Law of the Sea, the United

Breakwater in Entrance of Yongxing Island in Xisha Islands

Nation set up Preparatory Commission for the International Seabed Authority (ISA)/International Tribunal on the Law of the Sea (ITLOS). The United States refused to be a member of the Commission, holding that the issue of the regime of the Area cannot be settled within the Preparatory Commission. Under this circumstance, the United States asked for negotiations on the section of the Area in UNCLOS. This represented the interests of major developed countries in the west participating in the conferences of Preparatory Commission.

Added to the resistance of some western developed countries against the UNCLOS, the underestimation of the countries in formulating this convention also led to its later revision. Many articles of UNCLOS were adopted based on the prevalent shortage in supply in the international metal market in the 1970s and the prediction of the timing of commercial deep sea mining at that time. This all predicted that large-scale explorations of deep sea resources will be carried out in the foreseeable future. Yet since the adoption of UNCLOS, the trend that the international metal consumption market remained slack implied that despite of the huge potential, the commercial exploration of deep sea resources is unlikely to be conducted in a short period of time.

In 1989, to uphold the integrity and universality of the UNCLOS leaders of the Group of 77 voluntarily ordered to negotiate over the contents of Part XI in UNCLOS with developed countries so as for the convention to go into effect earlier. Western countries also agreed to this request.

Thus, it ushered in a series of informal consultations under the aegis of U.N. Secretary-General on pending issues relating tothe provisions on international seabed resources development in UNCLOS. Two rounds of such negotiations were initiated from 1990 to 1994 with 15 sessions of meeting being convened.

The negotiation process can be divided into two phases. The first phase was devoted to the identification of some states' concerns, the development of approach to be adopted in examining these issues and the search for solutions. During the negotiations in this phase nine issues causing difficulties and disputes such as costs of contracting parties; the Enterprise; decision-making; the Review Conference; transfer of technology; restriction of production; compensation fund; financial terms of contract; and environmental protection were identified. During the second phase more precision was given to the results reached so far; additional points were raised for consideration and participants directed their attention to an examination of consolidated texts embodying these solutions and on the procedure whereby they might be adopted. This phrase of negotiation was open to all delegations.

With the joint efforts of all participants, two documents were finally agreed through the informal negotiations: *The Agreement Relating to the Implementation of Part XI of UNCLOS Passed on December 10th 1982 (The Agreement of Implementation* for short) and *The Resolution of the Agreement Relating to the Implementation of Part XI of UNCLOS Passed on December 10th 1982*, thus completing the revision of Part XI of UNCLOS.

The Agreement of Implementation consists of 13 articles, mainly reiterating the uniformed nature of UNCLOS and reaffirming the passing of *The Agreement of Implementation* as the best way for all states' accession to UNCLOS.

The Agreement of Implementation has a binding force on those states which have already submitted their official confirmation and instrument of ratification to join UNCLOS. It has no binding force on those which have not acceded to UNCLOS.

The formulation of *The Agreement of Implementation*

rendered many articles in the part of the Area, Basic Conditions of Prospecting, Exploration and Exploitation, and Statute of the Enterprise no longer applicable or having to be supplemented. Since then, the system of international seabed development established in UNCLOS was changed.

Moreover, *The Agreement of Implementation* also revised or stipulated for some issues concerning review conferences, financial terms of contracts and financial committee.

The Agreement of Implementation was also a fundamental amendment to the components of UNCLOS. On the one hand, the amendment kept pace with changes of international situation, met the requirements of western countries such as the U.S., excluded the barriers for authorizing and acceding to the UNCLOS and upheld its universality and integrity. All of these are significant for establishing and maintaining a stable international maritime order. On the other hand, The *Agreement of Implementation* was the product of further concession to developed countries and greatly relieved their obligations. In 1994, the enactment of *The Agreement of Implementation* directly contributed to the UNCLOS' coming into effect on November 16th 1994.

The Legal Status of Deep Sea Genetic Resources

The UNCLOS excluded biological resources from the resources in the Area. It defines the resources in the Area in express terms: "resources" means all solid, liquid or gaseous mineral resources in situ in theArea at or beneath the seabed, including polymetallic nodules. This definition strictly limits the resources in the Area as mineral resources and obviously excludes biological resources from this scope.

The resources in the Area were not identified or listed in specific types in earlier discussions. The definition of "resources" in the Area was controversial at the beginning. At the 25th Session of the UN General Assembly, Resolution 2749 (XXV), *Declaration of Principles Governing the Seabed and the Ocean Floor, and the Subsoil Thereof, beyond the Limits of National Jurisdiction*, was passed which is the most important resolution during the formation and development of the regime of international seabed area. Its very first article reaffirms that "The seabed and ocean floor, and the subsoil thereof, beyond the limits of national jurisdiction (hereinafter referred to as the area), as well as the resources of the area, are the common heritage of mankind."

In the following discussions of United Nations Seabed Committee[1], parties concerned had some disagreements on whether to incorporate the seabed biological resources into the scope of "the common heritage of mankind".[2] Some states began to define the

---

1. In 1967, the 22nd United Nations General Assembly adopted a resolution that decided to establish an ad hoc committee to study the peaceful use of seabed ocean floor beyond the limits of national jurisdiction. In 1968, the 23rd United Nations General Assembly adopted a resolution that changed the ad hoc committee to "Committee of Peaceful Use of the Seabed beyond National Jurisdiction Ocean Floor 'Seabed Committee'". Seabed Committee did a lot preparatory work for UNCLOS . Article 133 of UNCLOS.
2. Frida M. Armas Pfirter, " The Management of Seabed Living Resources in 'the Area' under UNCLOS", Report presented at the Tenth Session of the ISA, p. 7.

Liwan 3-1, the 18-storey deep-sea oil and gas platform was going to South China Sea from the CNOOC dock in Qinghai Port.

"resources" in the Area that can be incorporated into the scope of "the common heritage of mankind".

In 1970, a British working paper held that the sedentary species of biological resources used for commercial development should be under the regime of the Area (international seabed area). In 1971, Malta proposed a working paper, the "Draft Ocean Space Treaty", maintaining that the international regime established in accordance with the international legal system shall conduct on behalf of the international community the management of biological and non-biological resources in International Ocean Space beyond the limits of national jurisdiction. A working paper submitted by Canada stressed that it was uncertain about whether the (new) international regime should only regulate the exploitation and exploration of mineral resources or should also include that of the seabed biological resources. It also stated in the paper that it does not think that there are any significant seabed biological resources in the Area (international seabed area). In 1972, in its report for discussion purposes the United Nations Seabed

Committee pointed out that "many delegations hold that the regime of the Area (international seabed area) should both include the seabed biological and non-biological resources." Nevertheless, some other delegations still contended that the regime of the Area only applies to the non-biological resources in the seabed.

During the Third United Nations Conference on the Law of the Sea, the issue of whether to incorporate seabed biological resources into the scope of the resources in the Area (international seabed area) was not much discussed about. This was mainly for three reasons.

First, states attached greater importance to the economic value of seabed mineral resources in recent years and relating discussions were more centered on mineral resources and relating prospecting system. Second, some states sought for establishing a similar international seabed regime for the biological resources in high seas. Many other states worried that this may undermine the traditional freedom that they enjoy in high seas. Therefore, they were opposed to incorporating any biological resources lying in international seabed into the definition of "resources" in the international seabed area. Third, at that time many states held that there were no biological resources of economic values in the Area including sedentary species.[1]

However recently with a large amount of biological communities being discovered at deep seabed, the legal status of deep sea genetic resources in the Area has become more prominent. Presently, the protection and sustainable utilization of marine biodiversity in the seas beyond the limits of national jurisdiction including the biological and genetic resources in the Area has

1. Frida M. Armas Pfirter, " The Management of Seabed Living Resources in 'the Area' under UNCLOS", Report presented at the Tenth Session of the ISA, p. 8.

become a hot issue in the field of international law of the sea.

The United Nations General Assembly established the United Nations Ad Hoc Open-ended Informal Working Group to Study Issues Relating to the Conservation and Sustainable Use of Marine Biological Diversity beyond Areas of National Jurisdiction (hereinafter referred to as the Ad Hoc Working Group for short) discussing this issue. There were fierce fights between developing countries and developed countries with regard to the legal status of marine genetic resources in the sea areas beyond the limits of national jurisdiction. Developing countries maintained the regime of Area of UNCLOS and proposed to define the marine genetic resources in the sea areas beyond the limits of national jurisdiction as the common heritage of mankind. In contrast, developed countries insisted on applying the regime of high seas of UNCLOS and claimed that any state has the freedom to develop such resources. It is worth pointing out that the marine genetic resources in the sea areas beyond the limits of national jurisdiction include both the biological resources in the Area and in high seas.

In 2011, the Ad Hoc Working Group raised a suggestion on the issues relating to the conservation and sustainable use of marine biological diversity beyond areas of national jurisdiction requiring the United Nations to initiate a process to reach a package of agreements on the protection and sustainable use of marine biological diversity beyond areas of national jurisdiction especially the conservation, sustainable use and benefit-sharing of marine genetic resources, construction of marine protected zones, capacity building and marine technology transfer in the sea areas beyond the limits of national jurisdiction. This suggestion won the approval of the United Nations General Assembly that year. Relating negotiations are still underway.

MINING IN INTERNATIONAL SEABED AREA

Prospecting

"Prospecting" means the search for deposits of polymetallic nodules (polymetallic sulphides, cobalt-rich crusts) in the Area, including estimation of the composition, sizes and distributions of polymetallic nodule deposits and their economic values, without any exclusive rights.

There shall be no time limit on prospecting except that prospecting in a particular area shall cease upon written notification to the prospector by the Secretary-General that a plan of work for exploration has been approved with regard to that Area. Prospecting shall be conducted in accordance with UNCLOS and these Regulations and may commence only after the prospector has been informed by the Secretary-General that its notification has been recorded. Each notification shall be formulated in accordance with the format stipulated by the Authority and contain the following information such as the basic information of the prospector, coordinates of the area to conduct prospecting, general description of prospecting plan, letter of commitment and so on. In the letter of commitment, the proposed prospector should state in express terms that he or she will comply with the relevant rules, regulations and procedures of UNCLOS and the Authority concerning cooperation in the training program with regard to marine scientific research and technology transfer stipulated in UNCLOS, and the protection and preservation of the marine environment, accept verification by

Montipora Foliosa at the Sea Bed in Xisha Islands

the Authority of compliance therewith and provide relevant data about the protection and preservation of marine environment to the Authority.

Prospectors have two rights. On the one hand, a prospector may recover a reasonable quantity of minerals to be used for testing (instead of for commercial uses). On the other hand, a prospector may report his or her part of the costs of development before commercial production as upcoming prospecting expense. The prospector shall also perform the following three obligations: First, to protect and preserve the marine environment in the process prospecting; second, to cooperate with the Authority to implement the training program with regard to marine scientific research and technology transfer; third, to submit an annual report of prospecting to the Authority.

Prospecting shall not confer on the prospector any exclusive rights or rights with respect to resources. Other entities also have the rights to apply to prospect in the area in which prospecting is being conducted by the prospector. Once the application is approved by the Authority, the prospector shall stop his or her prospecting immediately.

Exploration

"Exploration" means searching for deposits of polymetallic nodules (polymetallic sulphides, cobalt-rich crusts) in the Area with exclusive rights, analyzing such deposits, testing the collecting systems and equipment, processing facilities and transportation systems, and carrying out studies on the environmental, technical, economic, commercial and other appropriate factors that must be taken into account in exploitation.

the 3000-meter Deep-water Semi-submersible Drilling Rig

Each applicant for exploration shall submit "an application for approval of a plan of work for exploration in the form of cooperation" (exploration application) and activities could only be conducted after he or she gains the approval of the Authority.

Each application for approval of a plan of work for exploration shall be in the form prescribed in Annex 2 of UNCLOS to these Regulations.

Information of the applicants. In the case of an entity, the application should be submitted with a certificate of the sponsorship.

Materials about the area in which exploration is to be carried out including the nautical charts and local coordinates of the limits of the applied blocks, and remarks about whether the applicants choose to provide reserved areas or some shares arranged by joint enterprises for the Enterprise, and so on.

Information of the financial and technical capabilities of the applicants. An application for approval of a plan of work for exploration submitted by a State or a state enterprise shall include a statement by the State or the sponsoring State certifying that the applicant has the necessary financial resources to meet the estimated costs of the proposed plan of work for exploration. An application for approval of a plan of work for exploration submitted by an entity shall include copies of its audited financial statements for the past three years. In terms of technological capability, all applications shall include a general description of the applicant's previous experience, knowledge, skills, technical qualifications and expertise relevant to the proposed plan of work for exploration, a general description of the equipment and methods expected to be used in carrying out the proposed plan of work for exploration and other relevant non-proprietary information about the characteristics of such technology and a general description of the applicant's financial and technical capability to respond to any incident or activity which causes serious harm to the marine environment.

Information of the schedule of exploration. A general description and a schedule of the proposed exploration program, including the program of activities for the immediate five-year period, a description of the program for oceanographic and environmental baseline studies, a preliminary assessment of the possible impact of the proposed exploration activities on the marine environment, a description of proposed measures for the prevention, reduction and control of pollution and other hazards, as well as possible impacts, to the marine environment and a schedule of anticipated yearly expenditures in respect of the program of activities for the immediate five-year period should be submitted.

Undertakings. Each applicant shall provide a written undertaking to the Authority claiming that it will accept UNCLOS

In August 2013, the "Mona Princess" vessel loading jacket of Panyu "10-2", the jointly developed engineering project of Enping Oil Fields was sailing to the South China Sea.

as enforceable and comply with the applicable obligations created by the provisions of the UNCLOS and the rules, regulations and procedures of the Authority, the decisions of the relevant organs of the Authority and the terms of its contracts with the Authority; accept the control by the Authority over activities in the Area, as authorized by the UNCLOS and provide the Authority with a written assurance that its obligations under the contract will be fulfilled in good faith.

General-Secretary will inform the Legal and Technical Commission after receiving the undertakings. The written undertaking would be deliberated on and reviewed by the LTC and a written recommendation would be submitted to the Council which would make a final decision as to whether the undertaking should be approved.

After a plan of work for exploration is approved by the Council, the applicants shall sign an exploring contract with the Authority. After signing the contract, the proposed explorer will become a contractor. The duration of the contract is 15 years. The contractor shall have the exclusive right to explore the area covered by a plan of work for exploration. The Authority shall ensure that no other entity operates in the same area for the same resource in a manner that might interfere with the operations of the contractor. When engaging in the operations related with other resources, other entities should adopt the manners of operation that do not interfere with the operations of the contractor. A contractor who has an approved plan of work for exploration shall have a preference and a priority among applicants submitting plans of work for exploitation for commercial uses of the same area and resources in the future. On the expiration day, the contractor could

Shanghai Donghai Bridge Offshore Wind Farm

choose a place in his exploration area for commercial exploitation.

Contractors could report the exploration costs actually and directly used in exploration activities as their part of the development cost before commercial production. The obligations of the contractors include training personnel from the Authority and developing countries, performing waiver of obligations according to preset schedule, carrying out the plan of work approved by the Authority and preventing, reducing and controlling marine pollution and damages caused by the activities and so on.

Exploitation

"Exploitation" means the recovery for commercial purposes of polymetallic nodules (polymetallic sulphides, cobalt-rich crusts) in the Area and the extraction of minerals therefrom, including the construction and operation of mining, processing and transportation systems, for the production and marketing of metals.

After the expiration day of the contractor, contractors shall submit applications to the Authority, asking for the approval of a plan of work for exploitation in the Area if they want to conduct commercial exploitation. A contract shall be signed. One thing deserving special mention is that entities empowered to conduct exploration in the Area are not restricted to the exploration contractors only.

According to the regulations, "a contractor who has an approved plan of work for exploration only shall have a preference and a priority among applicants submitting plans of work for exploitation of the same area and resources", which means that when a contractor of a certain exploration area is unwilling or unable to exploit resources of this area, other entities conforming to the provisions of UNCLOS can apply to the Authority for the rights to explore related resources.

In UNCLOS, there are some articles about the policies on exploiting polymetallic nodules like production quota and

the Picturesque Scenery of East Lake in Haikou, Hainan Province

fee system, but due to the violation of market and commercial principles, many of those provisions were revised or abandoned by the 1994 Agreement mentioned above. According to Annex Section 6 of *The Agreement of Implementation*, the production policy on the Authority in the Area shall be based on the following principles:

(a) Development of the resources of the Area shall take place in accordance with sound commercial principles;

(b) The provisions of the *General Agreement on Tariffs and Trade*, its relevant codes and successor or superseding agreements shall apply with respect to activities in the Area;

(c) In particular, there shall be no subsidization of activities in the Area except as may be permitted under the agreements referred to in subparagraph (b);

(d) There shall be no discrimination between minerals derived from the Area and from other sources. There shall be no preferential access to markets for such minerals or for imports of commodities produced from such minerals;

(e) The plan of work for exploitation approved by the Authority in respect of each mining area shall indicate an anticipated production schedule which shall include the estimated maximum amounts of minerals that would be produced per year under the plan of work;

(f) The following shall apply to the settlement of disputes concerning the provisions of the agreements referred to in subparagraph (b): Where the States Parties concerned are parties to such agreements, they shall have recourse to the dispute settlement procedures of those agreements. Where one or more of the States Parties concerned are not parties to such agreements, they shall have recourse to the dispute settlement procedures set out in UNCLOS;

Currently, the Authority is formulating the exploitation regulations of polymetallic nodules to assure the rights and obligations of the exploiter and promote and standardize the commercial exploitation in the Area.

GOING TO THE OCEAN IS AN INEVITABLE COURSE FOR CHINA

Inborn Disadvantages

China is located in East Asia, west shore of the Pacific Ocean. With regard to territorial constitution, it is both a continental and maritime nation. First it is counted as a continental nation, covering a land territory of approximately 9.6 million square kilometers with land borders of 22,800 kilometers long.[1] The distance from east to west measures about 5,200 kilometers, and from north to south, 5,500 kilometers. Meanwhile, China is a maritime nation. Mainland China is surrounded to the east and south by the the Bohai Sea, the Yellow Sea, the East China Sea and the South China Sea. The mainland coastlines stretch from the north at the mouth of Yalu River on the China-North Korea Border to the south at the mouth of Beilun River on the China-Vietnam Border, measuring about 18,000 kilometers.

China has more than 7,300 islands with an exposed area out of water larger than 500 square meters. Including the mainland and islands area, the sea area China can claim its jurisdiction over has reached to 3 million square kilometers, according to relating stipulations in some international laws like the UNCLOS.

1. Li Yihu (2008): *A Re-examination of the Relations between Land and Sea in China: From Dichotomy to Coordination. Contemporary International Relations* (08), P1-7.

Sansha Municipal People's Government in Yongxing Island in Xisha Islands in Hainan Province

However, China's surrounding sea areas, especially the South China Sea, is one of the world's regions with the largest amount of disputes over islands, the acutest issues of maritime delimitation, the fiercest resource competition, and the most complicated geopolitical situation. Over half of the 3 million-square-kilometer sea areas under the jurisdiction of China overlaps and has disputes with the sea areas that the neighboring countries claim jurisdiction over.

As both a continental and maritime nation, the natural and geographical environment of China's ocean is characterized by the following points:

Even if China has the third longest coastline in the world, the per unit area of coastline is very short considering the large size of its land territory. The coastline coefficient of China is only 0.00188, which rates the 94th in the world. This number is lower than the world's most coastal states, which indicates that it is not convenient for the people in the vast inland areas in China to make use of the ocean.

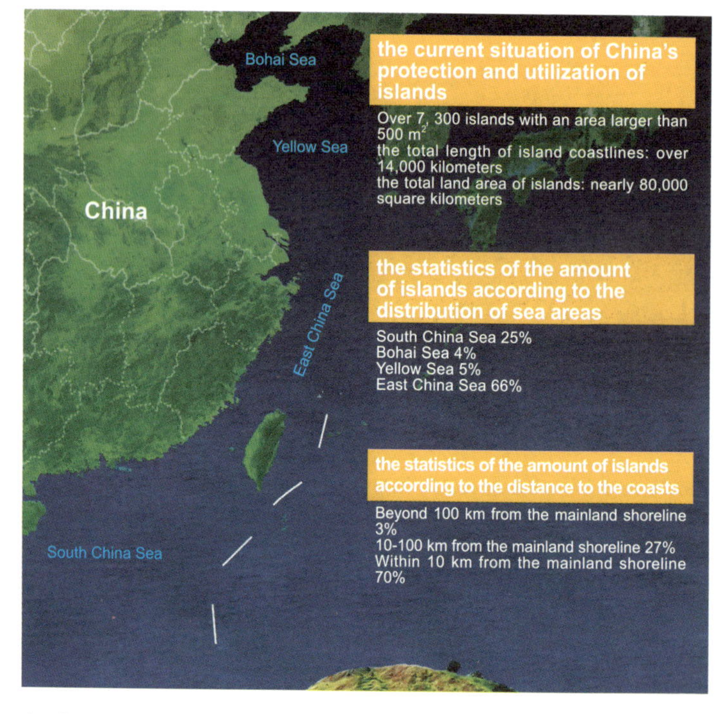

the Current Situation of China's Protection and Utilization of Islands in 2012

The sea areas China can claim over under its jurisdiction are 3 million square kilometers, but the size of the ocean is only 3/10 of the land, which is much lower than the average level of the world. According to UNCLOS, the total area of sea areas that can be counted within the jurisdiction of coastal states is about 109 million square kilometers, land area 136 million square kilometers (Antarctica not included). The sea/land area ratio is 0.96, the world's average level much higher than the ration of China.[1]

1. China Ocean Development Report Team: *China Ocean Development Report 2011*, China Ocean Press, P21-25.

China is surrounded by four seas in the east and blocked by island chains beyond the seas. The natural and geographical conditions are very unfavorable. The Bohai Sea, the Yellow Sea, the East China Sea and the South China Sea, marginal seas to the Pacific Ocean, east of mainland China are surrounding China in the shape of an arc with a large radius. No land but the Taiwan Island is directly facing the ocean, which is another natural disadvantage. According to UNCLOS, "enclosed or semi-enclosed sea" means a gulf, basin or sea surrounded by two or more States and connected to another sea or the ocean by a narrow outlet or consisting entirely or primarily of the territorial seas and exclusive economic zones of two or more coastal States. According to such definition, the Bohai Sea, the East China Sea and the South China Sea are all enclosed or semi-enclosed seas.[1] China has to go through many straits and water channels before finally going it to the sea.

China has a huge population. The amount of natural resources per capita is relatively small. The neighboring seas are all enclosed or semi-enclosed seas. Its resource and environmental carrying capacity is quite limited. The geographical conditions of its seas are quite unfavorable. Restricted by geographical and geological conditions, China's outer continental shelf extending beyond is also very limited. As a space for China's future strategic development and one of the sources of its strategic resources, the important and positive significance of the Area is self-evident. Thus, going to the ocean is an inevitable course for China's ocean development.

1. Although there is only one narrow outlet connecting to another sea, the Bohai Sea is completely encircled by China's coasts, thus it is neither enclosed nor semi-enclosed sea.

Important Strategic Resource Base

The UNCLOS adopted in 1982 regarded the international seabed area as the "Area", referring to the seabed, ocean floor and subsoil beyond national jurisdiction. The Chinese ocean activities usually referred to are mainly about the relevant activities in this "Area."

The Area abounds in diverse resources. The resources are specially stipulated in UNCLOS refer to all solid, liquid or gaseous mineral resources in situ in the Area at or beneath the seabed, including polymetallic nodules.[1] Polymetallic nodule is one of the

China's first deepwater pipe-laying crane ship *Hai Yang Shi You 201* was slowly leaving the dock of CNOOC at the Port of Qingdao pulled by 2 tugboats for the South China Sea in May 2012.

1. Article 133 of UNCLOS Article 133

main resources that human beings have discovered in the Area and also one of the earliest discovered resources of the Area.

So far, the deep-sea strategic resources of potential commercial and mining values that have been discovered in the Area include polymetallic nodules, cobalt-rich crusts, natural gas hydrates, hydrothermal sulphides, and biological genes that can survive under extreme conditions and so on. It is estimated that there are about 3 trillion polymetallic nodules and about 1 billion cobalt in the international seabed area. The amount of resources with the prospects for commercial exploration in the Clarion-Clipperton Fracture Zone (CCFZ) in the Northeast Pacific Ocean has already achieved 75 billion tons. Deep-sea genetic resources are new type of resources attracting world's attention in recent years, which have already brought billion dollars of industrial values.

Even though UNCLOS declares the Area as "common heritage of mankind", the new round of international "blue enclosure movement", characterized by taking and exploring resources in the Area, has already been underway. The exploitation and use of deep-sea resources has become a new heated issue for international competition. deep-sea resources has become a new international bid.

China is a big consumer of mineral resources, whose consumptions of aluminum, copper, iron, lead, tin and zinc occupies 20 percent of the world's total amount. The supply of strategic resources encounters a severe bottleneck. The reserves of some resources such as iron, gold, copper, oil, natural gas, aluminum bauxite are quite small and the conditions of their exploitation and utilization are far from favorable. Furthermore,

China's reliance on the import of important strategic resources such as oil and gas, cobalt and nickel is expanding. It faces a continuously imminent issue of the sustainable utilization and security of resources.

Exploiting resources in the Area plays a vitally important role for easing the conflict between demand and supply in China, reducing its dependence on import and ensuring the safety of resources supply.

Research Promotes Technological Development

Ocean is an important site and space for people to study the key frontier problems of marine science in the fields of deep-sea environment, deep-sea biology and the evolution of the earth. During the past nearly 50 years, the great breakthroughs made in the field of international earth sciences are mainly due to deep-sea research.

Plate tectonic theory proposed by the Canadian geophysicist John Tuzo Wilson is one of the most important scientific achievements of the 20th century. His view on global evolution is honored as a revolution in earth sciences causing a revolutionary change in the view of global evolution, no less striking than Darwin's theory of evolution. It provided a brand new scientific theory for people to recognize the formation and evolution of the earth's crust, the law of earthquake and volcano activities and the formation mechanism of minerals and fossil energy. The discovery

In 2011, China's scientific research vessel *Da Yang No.1* was in the middle of its 22nd scientific expedition and the TV grab was collecting and recovering samples from deep sea.

of deep-sea hydrothermal sulphides, the theory of the ocean circulation and the study on air-sea interaction (for example, the relation between the El Niño phenomenon and climate changes, the function of the ocean in global carbon cycle, etc) contributed quite a lot to exploration of major scientific problems concerning the survival and development of human beings such as the discovery of new deep-sea resources, the investigations of causes of global warming and the probing into the origin of life.

Marine science remains a frontier science of the 21st century. There will be more breakthroughs in theories with regard to the interaction and its process mechanism between various circle layers in the ocean, the coupling relation between global climate change and human vital activities, extreme marine conditions and deep biosphere, the deep geo-dynamic process and the evolution of continental margin and rock sphere in mid-ocean ridge. Therefore, many countries have put a premium on the basic scientific research on deep-sea areas and formulated related research plans. Actively conducting oceanic scientific research strongly promotes China's research level of earth sciences in deep-sea areas.

Exploration and exploitation of the Area exerts great importance on promoting China's technology of oceanographic observation, diving technology, technology of exploiting deep-sea mineral resources and enhancing the diversification and integration of its deep-sea high technology development.

The advanced deep-sea detection technology and equipment concern national security. Developed countries have long been implementing technology blockade against China. Under the support of the State High-Tech Development Plan (863 Program) and state special funds, China has ushered in rapid development of the R&D of equipment used for oceanographic investigations. China has independently designed and manufactured successively the 7,000-meter manned submersible, 6,000-meter AUV (autonomous underwater vehicle), 3,500-meter ROV (remotely operated underwater vehicle), deep-sea drilling rig, TV grab sampler, detection equipment of hydrothermal sulphide anomaly, deep-sea side scan sonar and sub bottom profiler. These have all made considerable contributions to China's development of deep-sea high technology.

In *An Outline of the National Medium and Long Term Program for Science and Technology Development* (2006 to 2020), China put in express terms "accelerate the development of aerospace and marine technology" on the list of the five strategic focus of the development of science and technology that China will concentrate on in the next 15 years. Ocean scientific research/ survey vessel has also been listed as one of the top priorities of future development.

Ocean resources survey vessels and equipments, as the research platform for the exploration of seabed resources and ocean detection are both important components of the marine capacity building and an embodiment of comprehensive national strength, whose development level directly influences a nation's marine technological potential. Actively conducting the exploration and exploitation of the resources in the Area contributes to the development of China's relevant industries and the upgrading and transformation of technology and equipment.

In general, UNCLOS encourages developing countries and geographically disadvantaged states to participate in activities in the Area. As a developing country with a quarter of the world's population, China should play an important role in the Area, share due benefits, and uphold the peaceful, reasonable and fair use of "common heritage of mankind". It is a necessary choice for China to greatly develop its maritime undertaking and comprehensively promote ocean scientific research so as to protect its interests in the Area and expand more space for its economic and social development.

大洋一号
坚决完成中国大洋26航次综合海

TWO DECADES OF CHINA'S MARITIME UNDERTAKING

Marching to the Three Oceans

China launched its exploration on international seabed in the middle of 1970s. Since then, China's oceanographic research vessels have marched into the ocean to carry on scientific research. Since 1980s, China's scientific and technological workers have landed on the continent of Antarctica for many times to conduct scientific research, and have pushed China's oceanic and scientific research undertaking into a new phase.

On March 30th 1976, according to the requirements of oceanic hydro-meteorological conditions to guarantee the success of the first long-range carrier rocket approved and forwarded by the State Council and Central Military Commission, State Oceanic Administration sent out scientific vessel including *Xiangyanghong 05*. They sailed from Guangzhou to specific marine areas in the central part of the Pacific Ocean to carry out comprehensive survey. This was the first oceanic research for China's scientific research vessels. As of October 21st, 1978, scientific research vessels including *Xiangyanghong 05* had carried out oceanic scientific research for four times in all, which lasted 265 days with a total voyage of over 130,000 kilometers.

During the four times' oceanic research, oceanic research vessels including *Xiangyanghong 05* traveled through Eastern and Western hemispheres and across the equator for many times,

In November 2011, CNPC pipe-laying vessel *Zhong You Hai 101* set sail from the city of Qingdao, Shandong Province, to Hong Kong for field operation.

overcoming numerous difficulties. Scientific and technological workers had acquired more than 100,000 pieces of meteorological elements and various atmospheric data on sea surface and more than 27, 000 pieces of hydrologic elemental data from sea surface to a depth of 1,600 meters (the deepest was 2,500). They had gathered water-depth measurement data of over 31,000 nautical miles, gravity data of over 27,000 nautical miles, and geomagnetic data of 3,100 nautical miles. They had collected 150 bottles of plankton samples and acquired massive data in the fields of oceanic communication test, deep-ocean under-water sound transmission test, sound velocity test, test for oceanographic instruments and so on.

The most important scientific finding among them was that, in the third oceanic research, scientific and technological workers had collected one geological samples of manganese nodule of the size of two fists' at a depth of 4,784 meters, in the sea area of 171° 30′ E, 6° S. This was the first piece of manganese nodule China obtained from the ocean. Its finding marked the start of China's poly metallic nodule research of polymetallic nodules in international seabed area. It also qualified China in applying to be a pioneer investor to United Nations Seabed Committee.

In the four successful rounds of ocean scientific research conducted by oceanic research vessels including *Xiangyanghong 05*, China not only acquired massive scientific data, but also gathered abundant experience in the aspect of PLAN oceanic voyage, ocean weather broadcast, long-distance encrypted communication, commanding and cooperative of vessels, responding to other countries' air and sea tracking interference, and so on, and trained a large group of middle- and senior-level professionals and commanders. The successful ocean scientific research of *Xiangyanghong 05* unveiled the prelude for China

The *vessel XiangYangHong 09* was sailing from Qingdao, Shandong Province, to Mariana Trench to challenge a 7,000-meter deep dive.

to march to the ocean and greatly pushed the development of its ocean research.

From December 1977 to November 1979, World Meteorological Organization implemented the "First Global Weather Experiment". It was the largest subplan to the "Global Atmospheric Research Program" (GARP) jointly organized by the World Meteorological Organization and International Council for Science (ICSU). More than 140 countries and regions took part in this international cooperation. Chinese government sent out two oceanographic research ships of *Shijian* and *Xiangyanghong 09* to carry out the task of tropical wind observations. The ships *Shijian* took 213 days to carry out two cruises of investigations and tests in the Western Pacific Area for two cruises from December 1978 to July 1979, covering a total voyage of 32,840 nautical miles. *Xiangyanghong 09* took 197 days to carry out two cruises of investigations and tests in two specific marine areas of the Western Pacific Area, covering a total voyage of 28,641 nautical miles.

In this international scientific cooperation, two Chinese oceanographic research ships completed the mission of observing high-altitude meteorology, sea surface meteorology and the temperature from sea surface to a depth of 200 meters. Besides, they established 4 sections in the investigated sea area, 38 and 42 observation stations respectively to meet the need of Institute of Oceanology, Chinese Academy of Sciences and carried out a comprehensive scientific investigation through the method of single-boat measuring tour. That was China's first participation in international cooperation in oceanographic scientific research and investigation. China has accumulated rich experience for its oceanographic scientific research to further promote its undertaking of oceanographic scientific investigation, enhance its academic communications of ocean sciences with world's organization of ocean sciences and other countries.

In October 1979, the large-scale comprehensive oceanographic scientific survey and ocean communication ship *Xiangyanghong 10* independently researched and manufactured by China for the first time was completed and delivered to use to East China Sea Branch of State Oceanic Administration. Initially, it was designed for the full flight test for China's long-range carrier-rock, whose missions mainly covered oceanographic investigation, ocean weather and communication. These tasks should be performed by three different types of ships: ocean survey ship, ocean weather vessel and dispatch vessel according to the international conventional classification of ships. However, due to lack of funds, Chinese designers combined the features of these three types of ships in one and built a large-scale comprehensive multi-purpose vessel. It was unprecedented at that time in the whole world. Japanese media reported it as the large-scale Special Ship.

After being put into service, the *Xiangyanghong 10* has executed ocean scientific survey tasks for many times. In April 1980, as a flagship, it participated in China's first test launch of carrier-rocket to the Pacific Ocean. In April 1984, it took part in took part in the launch of China's first test launch of synchronous fixed-point experimental communications satellite. From November 20th 1984 to April 10th 1985, it crossed the Pacific Ocean and carried out China's first scientific expedition of the Southern Ocean and Antarctica. In this investigation, it complete d comprehensive observations over 34 stations with a measuring line of 3,600 nautical miles and investigated 23 projects including marine biology, chemistry, meteorology, hydrology, geography, and geophysics. It obtained the first-hand data, samples and specimens of on-site observations, filling in the blank in the domestic related fields and made outstanding contributions to the development of China's ocean scientific expedition. In 1985, the vessel won the grand prize of Chinese first National Science and Technology Progress Award.

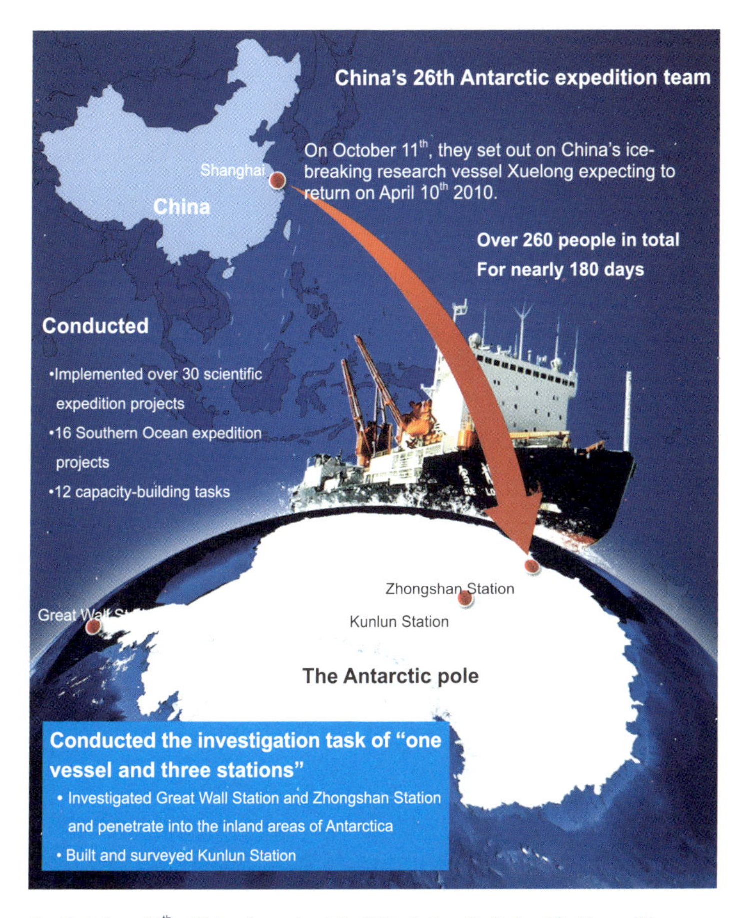

On October 11th, China launched its 26th Antarctic Scientific Expedition.

In the 1980s, China also carried out many ocean scientific investigations and surveys. These mainly included the following activities: In 1983, the ocean survey ship *Kexue 1* carried out geophysical investigation on the West Pacific. During this investigation, the ship completed the observations over the oceanic crust with seismic buoys which was unprecedented in China's history of oceanographic survey. In 1983, the ship *Xiangyanghong 16* conducted an investigation of manganese nodules in the west sea areas of North Pacific and obtained several kilograms of manganese nodule samples among which the largest weighed 2.9 kilograms. In October 1984, China formed its first Antarctic research expedition team marking its investigation over Antarctica. In February next year, China established its first scientific expedition station of Great Wall Station in Antarctica.

Since the 1990s, Chinese ocean scientific expedition ushered into a new stage. On April 9th 1990, with the approval of the State Council, China Ocean Mineral Resources R&D Association (COMRA) was established. On March 5th 1991, with the approval of the United Nations COMRA registered as the world's fifth international undersea development pioneer on the International Seabed Authority and the Preparatory Commission of the International Tribunal for the Law of the Sea, and was assigned 150,000 square kilometers of the development zone in the international seabed area beyond the national jurisdiction. The establishment of COMRA marked that China has marched into the ocean in an all-round way.

COMRA is a social organization organizing and coordinating its members in participating in the research and exploitation activities in the international seabed area. Its aims are exploring new resources for China through resource research and exploitation activities in the international seabed area, advancing the formation and development of China's deep-sea high-tech industry, upholding China's rights to explore international sea-bed resources and making contributions to the utilization and exploitation of international sea-bed resources of mankind. The highest policy-making body of COMRA is the Council composing of members. There are both unit members and individual members. Presently, there are 74 unit members. COMRA has organized and carried out its tasks with efficiently and fully fulfilled its coordinating function since its founding.

When COMRA was first established, it considered the investigation and evaluation of polymetallic nodule resources as the focus of its work. After its registration as an international seabed development pioneer, COMRA organized the scientific expedition vessels such as *Haiyang Sihao*, *Xiangyanghong 16*, and *Dayang Yihao* to conduct 10 cruises of investigation and survey in the applied sea areas.

On March 5th 1999, according to UNCLOS, China performed its obligation of waiver of 50% of its pioneer area and obtained the exclusive exploration right and mining concession priority over an area of 75,000 square kilometers in its pioneer area. In May 2011, COMRA signed a *15-year National Seabed Polymetallic Nodule Resources Exploration Contract* with the International Seabed Authority. It defined in the form of legal document that China obtained a mining contract area of polymetallic nodules covering 75,000 square kilometers and had exclusive exploration rights and mining concession priority in this area, which expanded the total reserves of China's strategic resources.

The vessel of *Dayang Yihao* set sail from Qingdao for China's first global investigation.

On April 2nd 2005, the vessel *Dayang Yihao* set sail from Tuandao Bay in Qingdao for China's first scientific expedition to the three oceans. The focus of this scientific expedition was to investigate the seafloor hydrothermal area and special pheno menon of life around it. There are many organisms living in the extreme environment in the seafloor hydrothermal area, which, as well as their genetic resources, have great scientific research value and economic value. The vessel *Dayang Yihao* set out of the Pacific in the east, entered the Atlantic in the west through the Panama Canal, reached the India Ocean via Cape of Good Hope, and returned to the Pacific through the Strait of Malacca. It took 297 days and travelled for a voyage of 43,230 nautical miles. Thus, China fulfilled its long-cherished dream of "marching to the three oceans".

On December 8th 2010, the vessel *Dayang Yihao* set sail again from Guangzhou, executed a task of scientific investigation in the Indian Ocean, the Atlantic and the Pacific. During the scientific investigation this time the vessel carried out various projects of survey and investigation, including polymetallic sulfide in the seabed, polymetallic nodules, deep-sea environment, deep-sea biological gene and deep-sea biodiversity and so on. That was a highly comprehensive ocean survey voyage. Scientific workers completed the task of scientific investigation successfully when they returned and achieved lots of historic breakthroughs: found 16 submarine hydrothermal areas, 5 from the Atlantic, 11 from the Pacific. Among them, the hydrothermal area discovered at latitude 26 degrees South in the South Atlantic is presently the southernmost hydrothermal area in South Mid-Atlantic Ridge. The vessel captured a suspected new species of deep-sea fish and a large number of hydrothermal creatures such as blind shrimp. It for the first time microbial membrane samples at different depths of water and genetic resource samples in different environment of

On March 26th, 2012, the ocean scientific research ship *Dayang Yihao* was about to embark on a new journey. It was going to cross many oceans and conduct the 26th ocean expedition. The picture shows that the ship would conduct research and investigation over seabed minerals, sulfides, and hydrothermal fluid with the underwater sulfide detectors it carried onboard.

Expedition members were observing and studying seabed sulfide collected by the vessel *Dayang Yihao*.

large spatial scale. It provided precious samples for the research of deep-sea microbial diversity and the acquisition of genetic resources. It for the first time obtained the large-scale, multi-station and full-depth environmental data and biological and microbiological samples in the area of particular concern of polymetallic nodules environment for the first time. This voyage lasted for about 369 days covering through 9 navigation sections and a distance of 64,162 nautical miles.

China started late for the survey of polymetallic sulfide resources in the ocean, but developed at a fast speed. Since 2006, COMRA has organized a number of scientific research activities and discovered nearly 20 hydrothermal areas on the seafloor of the Pacific, the India Ocean and along the Mid-Atlantic Ridge, which accounted for 10% of the global seafloor hydrothermal areas that have been discovered. In July 2011, the

17th session of International Seabed Authority was held in Kingston, Jamaica, approving the application of the plan of work of polymetallic sulfide exploration submitted by COMRA. China obtained the exclusive exploration rights over a polymetallic sulfide mining area of 10,000 square kilometers in the international seabed area in Southwest Indian Ocean and would enjoy mining concession priority in the future exploitation of this resource. That was the first application of mining area that the International Seabed Authority accepted and approved after the promulgation of *Regulations on Prospecting and Exploration for Polymetallic Sulphides in the Area* on May 7th 2010. The applied area is located in the Southwest Indian Ridge, within a limited rectangle area with a length of 990 km and a width of 290 km.

In July 2012, COMRA submitted an application of the exploration of "Area" cobalt-rich crusts to ISA. On July 19th 2013, the International Seabed Authority approved of the COMRDA application. China obtained an exploration contract area of 30,000 square kilometers of cobalt-rich crusts in the Western Pacific, making it the world's first country owning exclusive exploration mines of three major international seabed mineral resources in the world.

On May 28th 2014, China's scientific research vessel, the *Haiyang Liuhao*, set sail from the special purpose jetty of marine geology at the Pearl River Estuary to the Pacific Ocean to carry out scientific expedition and observation of deep sea resources and the ocean. It is China's 32nd ocean voyage task. On June 25th, within less than 17 days, the *Haiyang Liuhao* obtained 30 columnar samples of abyssal sediments, completed the CTD measurement of 3 stations and measuring line operation of 1,408 kilometers. It successfully

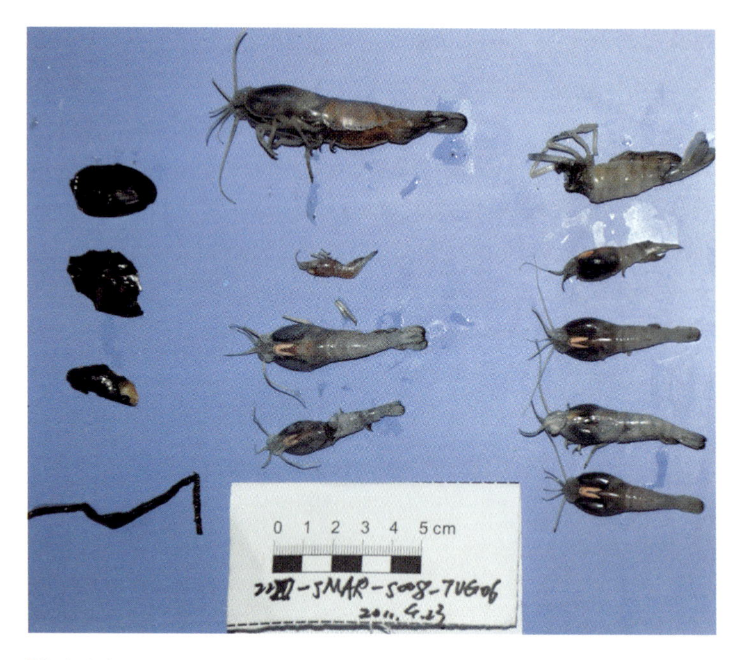

Blind shrimp specimens collected by the vessel *Dayang Yihao* in deep-sea sulfide areas.

the Deep-sea Sagefish at a Depth of 5,000 Meters

performed the voyage investigation of the deep-sea resources for China Geological Survey. That was China's first time for China to conduct sampling collection of abyssal sediments so systematically in this area, which will lay a solid foundation for the investigation and evaluation of resources in the future. On August 20th, after receiving supplies and equipment debugging in Honolulu in Hawaii, it set sail to the work site of nodules in east Pacific, beginning the scientific research of the second leg of the 32nd ocean voyage.

Up to 2014, China's State Oceanic Administration and the Ministry of Land and Resources had carried out ocean investigations and surveys dispatched ocean scientific research vessels which conducted ocean surveys for 32 cruises. They navigated three oceans and obtained precious investigation data providing important basic materials for resource assessment and research of environmental sciences.

From *Qianlong, Hailong* to *Jiaolong*

After more than 20 years of development, China has basically constructed a technological system of the exploration of deep-sea resources of the world's advanced level, formed the technological modeling of deep-sea mining system with independent intellectual property rights, and carried out the dressing and smelting processing technology of deep-sea resources with independent intellectual property rights.

After many rounds of modernized installations and retrofitting, China's ocean investigation vessels for special use have already transformed into a technological platform of ocean mineral resources exploration of the world's advanced level and basically kept pace and integrated with the development of international level in terms of deep-sea detection technologies and methods. China independently researched and manufactured a series of deep-sea exploration and investigation instruments such as deep-sea shallow drilling, deposit-grab with TV system and 6,000-meter acoustic towed system and so on and significantly improved its technological capability of deep-sea investigations. It has successfully researched and manufactured 6,000-meter AUV (autonomous underwater vehicle), 3500-meter *Hai Long* ROV (remotely operated underwater vehicle), and deep manned submersible *Jiao Long*.

Qianlong No.1

The *Qianlong No.1* is an autonomous underwater vehicle developed independently by China, serving in the resource exploration in the deep ocean. As one of the major projects in *China's Twelfth Five-year Guideline of Research and Development of International Marine Resources* it was developed and manufactured jointly by Shenyang Institute of Automation of the Chinese Academy of Sciences, commissioned by COMRDA, along with the Institution of Acoustics of Chinese Academy of Sciences and Harbin Engineering University commissioned by COMRDA. It was officially launched in November 2011. It completed its lake trial and test in March, 2013. The *Qianlong No.1*, a rotor with 4.6 meters long, 0.8 meters in diameter, weighs 1,500 kilograms. Its maximum working depth is 6,000 meters. Its cruising speed is two knots and its maximum endurance is 24 days. It is equipped with shallow seismic profile instrument and other detection equipment. It can complete missions such as a detailed detection of micro-topography of the seabed, detection of substratum, measurement of seabed hydrological parameters and richness index of seabed polymetallic nodules.

In May 2013, the *Qianlong No.1* onboard the scientific research vessel *Haiyang Liuhao* set sail for its initial trial in the South China sea. During the trial, the *Qianlong No.1* conducted 7 dives in total, with a maximum depth of 4,159 meters, obtaining a batch of detection data of the submarine topography. The rate of equipment deployment and the recovery was 100%. This marked China's significant progress in the field of the practical retrofitting of deep-sea resource exploration equipment, guaranteeing its deep-sea resource exploration.

In May 2013, during the 29th ocean voyage research, the first practicality verification test of the *Qianlong No.1* was conducted. On October 6th, it was launched into the water in the designated areas to descend to a depth of over 5,000 meters and fulfill the preset missions such as acoustic detection with a fixed height of 50 meters. After 8 hours and 5 minutes, it sent out messages of floating up to the surface monitoring system on the water and it was retracted to the deck successfully at 19:30. As checked, the quality of the acoustic data obtained by the *Qianlong No.1* in the first dive was of good quality. The ocean experimental application of the *Qianlong No.1* gained initial success. At 20:00 on October 8th and 22:00 on October 9th, it conducted nightly diving operations twice, working under the water for 8.5 hours and 10 hours respectively, with all performance indexes of its equipment remaining stable.

On August 20th 2014, the vessel of *Haiyang Liuhao* carried out the 32nd voyage task of scientific research. The *Qianlong No.1*

the Scientific Research Ship of *Haiyang Liuhao*

again boarded on the vessel to carry out its second experimental application. In the China's contract area of poly-metallic nodules in the East Pacific, the *Qianlong No.1* carried out the diving operation successfully. At this time, it was launched into the sea to a depth of 5,151 meters and conducted underwater operation for six hours. According to the data after it was recovered and checked, all equipment in the underwater vehicle functioned normally and a large number of sounding data was obtained.

Hailong

Hailong, a national project of science and technology, is a powerful remotely operated vehicle (ROV) designed by the Underwater Engineering Institute of Shanghai Jiaotong University. It was then the underwater robot with the deepest diving meters and best working ability. In May 2008, *Hailong* completed a deep-sea test of 3,278 meters in South China Sea, achieving China's greatest diving record among the ROVs. *Hailong* is 3.17 meters long, 1.81 meters wide, 2.24 meters high in size, weighing 3.45 tons in the air (including the effective loading). Equipped with 5 multifunctional cameras and 1 still camera, and installed with 6 floodlights and 2 high brightness HID lamps, it can provide clearer picture quality and rich and detailed first-hand video data for marine research.

In July 2009, *Hailong*, in its twenty-first voyage to the East Pacific Ocean, observed a huge rare "black smoker" (the black smoke-shaped hydrothermal jet in the high temperature hydrothermal activity in the deep sea with rich sulfide) and accurately captured polymetalli c sulphide samples with its mechanical hand, completing China's first precise fixed-point sampling and observation of deep-sea hydrothermal chimney and obtaining valuable information. Deputy Director of General Office of COMRA remarked that was a historic moment and the successful field using of the most sophisticated technology and equipment for China's ocean survey, marking that China has become one of the few countries that can conduct investigation and sampling of hydrothermal along mid-ocean ridge with underwater robot.

In January 2014, *Hailong*, in its 30th ocean voyage carried out underwater operation for five times in the polymetallic contract area in Southwest Indian Ocean, four of which succeeded. During the process, *Hailong* gave its feature of precise and detailed investigation into full play, providing guarantee equipment for China in finding out the situation of the polymetallic sulfide exploration contract area. In the last underwater operation, *Hailong*, not only observed carbonate "smoker", but also successfully obtained a tube of water sample. The technology of the high precise positioning, real-time precise observation, picturing and sampling capability manipulated in this operation were hardly achievable with conventional methods. In operation, *Hailong* also broke through the record of near-bottom operations for 8 hours continuously, while the previous underwater robots could only work for three or four hours basically.

Jiaolong

To accelerate the development of deep-sea vehicle and exploration technology, China's Ministry of Science and Technology of the People's Republic of China listed deep-sea research submersible on the priority of State High-Tech Development Plan (863 Program). The research and development of *Jiaolong* was specifically organized and implemented by COMRA. Hundreds of research institutions and enterprises such as 702nd Research Institute of CSIC (China Shipbuilding Industry Corporation), Shenyang Institution of Automation and Institute of Acoustics of Chinese Academy of Sciences also participated in the technological brainstorm. *Jiaolong* is the first operational deep-sea manned submersible designed and manufactured independently by China. Its design maximum diving depth is at the 7,000-meter class. It is the operational deep-sea manned submersible that can dive to the deepest in the world at present. Its operation range can cover 99.8% of the world's ocean, which is of great significance for China's development and utilization of its resources. *Jiaolong* is 8.2 meters long, 3.0 meters wide and 3.4 meters high. Its bare weight is less than 22 tons and its maximum load is 240 kilograms. Its maximum design speed is 25 nautical miles per hour. Its cruising speed is 1 nautical mile per hour and its maximum design operation depth designed is 7,000 meters.

Jiaolong and Its Mother Ship *Xiangyanghong 09*

From May to July in 2010, *Jiaolong* had carried out the diving tasks for many times in the South China Sea, with a the maximum diving depth of 3,759 meters. On July 26th 2011, *Jiaolong* conducted its second diving test with a diving depth of which was 5,057 meters. It was the first time for it to break the record of 5,000 meters. From June 15th to 30th 2012, *Jiaolong* completed 6 dives in the Mariana Trench and reached a depth of over 7,000 meters for three times. In its fifth dive on June 27th, it reached the largest submerged depth of 7,062.68 meters.

On July 31st 2014, during China's 35th ocean voyage, *Jiaolong* conducted its dive in the first leg of this in the area of Caiwei seamount of the exploration contract area of cobalt-rich crusts of COMRA, the engineering diving. This was the 75th dive for *Jiaolong*. The maximum diving depth of this dive was 2,424 meters and the diving location was on the top of the southern slope of the seamount. Its main task included the near bottom navigation, observation, shooting the HD video of submarine megabenthos and cobalt rich crust, acquisition the samples of megabenthos, cobalt-rich crusts, rocks and sediments and near bottom water, shooting the process of submersible placement, diving, sitting on the sea bed, floating, recovering in the manned cabin, and carrying out the physiological and psychological tests for aquanauts. Three aquanauts conducted operation on the seabed area for nearly ten hours, and completed all scheduled work successfully, marking a successful beginning.

Up to now, there have been eleven aquanauts who dived at and over the depth of 7,000 meters under the water in the world's history of manned deep diving among which 8 are Chinese. The successful diving test of *Jiaolong* diving test marks China's ability to take on manned deep-sea exploration of more than 99% of the depths of world's sea areas. It also marks that China has reached the international leading level in manned undersea scientific research and resources exploration, , ranking it among the world's deep-sea club.

Manned deep-sea submersible *Jiaolong* was craning to be stalled to its mother ship *Xiangyanghong 09*.

Jiaolong launched its second experimental voyage

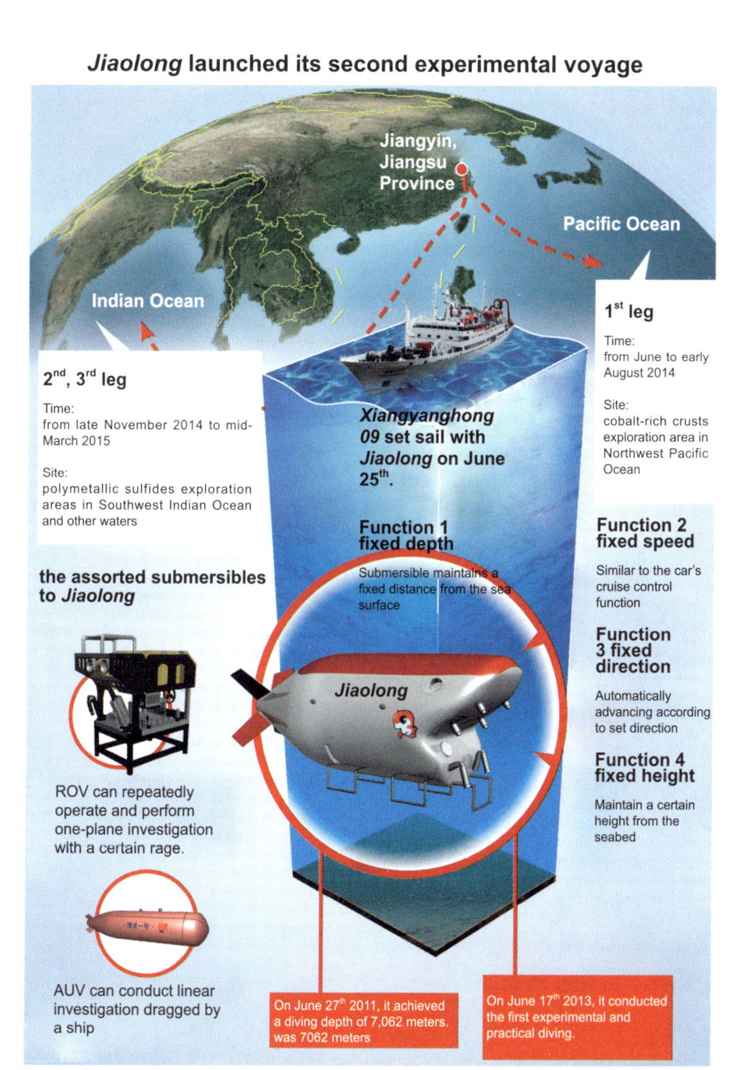

On June 25th, 2014, manned deep-sea submersible *Jiaolong* and its mother ship *Xiangyanghong 09* set sail from Jiangyin, Jiangsu Province to conduct its 40-day expedition task in Northwest Pacific Ocean.

Deep-sea R&D and Participation in Management

China has gained great achievements in deep-sea research and development. It has built the comprehensive microbial resource database and preserved over 4,000 microbial strains from the deep-sea area. Meanwhile, it has constructed the first large-fragment metagenomic library of deep-sea sediments and preserved over 50,000 large-fragment clones. Over 100 new metabolites were discovered, which laid a sound foundation for cultivating strategic emerging industries.

COMRA formulated and implemented the NaVaBa Plan which aimed to study deep-sea ecosystem and its annual change. Many researches about biology, chemistry, hydrology and geological baseline of different levels were conducted under the guidance of this plan. A relatively systematic database of biological, chemical, physical and geological environment of China's pioneer areas was constructed. NaVaBa Plan is one of China's great contributions to the environmental assessment and research activities carried out in the Area, which has largely enhanced China's international prestige.[1]

China attaches great importance to the function of the Authority and actively participates in its work. China also has

---

1. *Two Decades of China's Marine Undertaking*, *Ocean Development and Management*, 2011 (10).

established permanent representative offices in Jamaica, where the headquarters of ISA is located. In1996, China became a member of Group B at the first session of the Council of the Authority. In 2004, China was listed in Group A as the largest consumer of minerals in the Area. Chinese government donated to the Voluntary Trust Fund for four times (in 2008, 2009, 2012 and 2013) to make its own contributions. Besides, China also takes an active part in related international cooperation.

China has played a significant role in the codification of deep-sea mining act in the Area. The idea of formulating *The Regulations on Prospecting and Exploration for Polymetallic Sulphides and Cobalt-rich Crusts in the Area* was first proposed by the Russian delegation at the 4th session of the Authority in 1998. At the request of Russia, the Authority held in June 2000 a seminar with a theme of "other resources in the Area and in deep-sea areas besides polymetallic nodules". According to the results of the seminar, the Secretariat of the Authority compiled the document of *Considerations on The Regulations on Prospecting and Exploration for Polymetallic Sulphides and Cobalt-rich Crusts in the Area* before the 7th session of the Authority was convened, laying a foundation for the future formulation of regulation drafts. With the deepening of people's understanding of the two resources of polymetallic sulphides and cobalt-rich crusts, the Council of the Authority reached a consensus in formulating draft regulations on these two resources at the 12th session.

Since the discussion on formulating regulations on prospecting and exploration for polymetallic sulphides and cobalt-rich crusts at the 7th session of the Authority in 2001, China has for many times expressed its own positions at the sessions of the Authority and proposed its own ideas and suggestions with regard to the formulation of draft regulation. During the process of discussion of draft regulation, China on many occasions stated its support for the Authority to conduct the management of these two resources and meanwhile maintained that the work of regulations on prospecting and exploration should be carried out gradually without taking any random actions.

China pointed out that polymetallic sulphides and cobalt-rich crusts both exist in the international seabed area (the Area) and areas under national jurisdiction. Therefore, regulations on prospecting and exploration for polymetallic sulphides and cobalt-rich crusts in the Area should be competitive and an exploitation system suitable to these two resources should be explored and formulated. Besides, new prospecting and exploration system should fully concern the interests of the International Seabed Authority and of the developing countries in order to avoid any monopoly of a few developed countries on the prospecting and exploration of these two resources. China held that environmental protection related with the prospecting and utilization of new resources is of vital importance. There is a huge difference in environmental impact of crusts and sulphides and of nodules. The ecosystem and biological community of seamounts and ocean ridges differ from that of deep-sea basin in which nodules abound. The distribution of the two resources is relatively concentrated. The future prospecting and exploration will be carried out in a rather limited scope. These facts all require a strengthened evaluation and monitoring of the environment. Moreover, taking into consideration of the changes of the international environmental law since the passing of UNCLOS it is necessary to highlight and enhance the protection of marine environment and sustainable utilization of resources. China hoped that the Authority could play an active role in the protection of deep-sea environment and biodiversity through the formulation of the regulation and the establishment of a package of assorted systems. With the joint efforts of China and other countries, the International Seabed Authority passed *The Regulations on Prospecting and Exploration for Polymetallic Sulphides and Cobalt-rich Crusts in the Area* in May 2010.

On July 26th 2012, the session of the Authority reached a consensus to adopt the *Regulations on Prospecting and Exploration for Cobalt-Rich Ferromanganese Crusts in the Area.*[1] This is the third set of deep-sea mining act after the regulations on polymetallic sulphides nodules and on sulphides. The enactment of this Regulation was a consensus driven vigorously by the Chinese delegation.

At the 16th session of the Authority, the Council initiated the review on the regulations on prospecting and exploration for cobalt-rich ferromanganese crusts after adopting the *Regulations on Prospecting and Exploration for Sulphides.* However, since time was limited back then all parties only exchanged their ideas for a while. The Council required a further revision of the draft regulations for the regulations on cobalt-rich crusts in order to keep in line with the *Regulations on Prospecting and Exploration for Sulphides.* Chinese delegation expressed its concern for the undersized of the mining area in the draft regulations and proposed an informal suggestion of revising the articles concerning the size in the draft regulation. However, this suggestion was not discussed in the session.

During the 17th session of the Authority held in 2011, the Council continued the discussion of the articles relating with cobalt-rich crusts in the draft regulations and achieved

1. *Regulations on Prospecting and Exploration for Cobalt-rich Ferromanganese Crusts in the Area*, ISBA/18/C/L.3

an agreement on the most part of the draft. Chinese delegation suggested increasing the prospecting area to 3,000 square kilometers instead of 2,000 square kilometers and the exploration area to 1,000 kilometers rather than 550 kilometers. In this case, the area of cobalt-rich ferromanganese crusts block would be changed from "no more than 50 kilometers by 550 kilometers" to "less than 300,000 square kilometers rectangle with the longest length of side less than 1,000 kilometers"[1], but no agreement was reached in the session with regard to this part.

At the 18th session of the Authority held in 2012, the Council continued its discussion on the unsettled issues with regard to the draft regulation on cobalt-rich crusts. Finally, the Council accepted China's suggestion of revision on the issue of prospecting area, changing it to 30,000 square kilometers from 2,000 square kilometers and exploration area 1,000 square kilometers from 500 square kilometers. In terms of the scope of cobalt-rich crusts blocks, the Council decided to keep the original article of "no less than 550 kilometers by 550 kilometers" standard. With problems almost solved, *The Regulations on Prospecting and Exploration for Polymetallic Sulphides in the Area* was eventually approved of.

The *Regulations on Prospecting and Exploration for Cobalt-Rich Ferromanganese Crusts in the Area* is identical to the *Regulations on Polymetallic Sulphides* in terms of style. It is composed of ten parts and four annexes. Just like the *Regulations on Polymetallic Sulphides*, the *Regulations* changed the participation method of the Authority, adopting new concepts of "blocks", "clusters" and "contiguous"and others.

---

1. Permanent Mission of P.R.C to ISA: International Seabed Information, 2011 (43)

The Main Contents of the Three Regulations

	Regulations on Polymetallic Nodules	**Regulations on Polymetallic Sulphides**	**Regulations on Cobalt-Rich Ferromanganese Crusts**
Time of Approval	2000	2010	2012
Application Area	150,000 km^2	10,000 km^2	3,000 km^2
Exploration Area	75,000 km^2	2,500 km^2	1,000 km^2
Exploration System	Reserved Area	Reserved Area / Joint Venture	Reserved Area / Joint Venture
Fee for Application	$500000	$500000	$500000

According to *Regulations on Prospecting and Exploration of Cobalt-rich Crusts* the exploration and prospecting area that the applicants could obtain is no more than 3,000 square kilometers at the most. The area covered by each application for approval of a plan of work for exploration for cobalt-rich crusts shall be comprised of no more than 150 cobalt-rich crust blocks which shall be arranged by the applicant in clusters. Five contiguous cobalt-rich crust blocks form a cluster of cobalt crust blocks. Two such blocks that cross at any point shall be considered to be contiguous. Clusters of cobalt-rich crust blocks need not be contiguous but shall be proximate and located entirely within a geographical area measuring no more than 550 kilometers. The contractor shall fulfill his or her obligation of relinquishment during the prospecting. When the contract expires, the area for cobalt-rich ferromanganese crusts exploration and prospecting that the contract could reserve should be no larger than an area of 1,000 square kilometers.

Besides the environmental protection which is similar to the regulations on polymetallic sulphides, the *Regulations on Prospecting and Exploration for Cobalt-Rich Ferromanganese Crusts* emphasizes the protection of marine ecosystem with regard to seamounts and cold water corals. The Legal and Technical Commission (LTC) shall develop and implement procedures for determining, on the basis of the best available scientific and technical information, whether proposed exploration activities in the Area would have serious harmful effects on vulnerable marine ecosystems, in particular seamounts and cold water corals, and ensure that, if it is determined that certain proposed exploration activities would have serious harmful effects on vulnerable marine ecosystems, those activities are managed to prevent such effects or not authorized to proceed.

While actively participating in the discussions about the international seabed area, China has continuously enhanced the management of the domestic maritime undertaking. One of its significant measures is the formulation of law of the sea. During the first session of National People's Congress held in March 2013, 31deputies submitted a resolution on formulating the act on the prospecting and exploration of resources in the ocean. Presently, the has already been incorporated into the legislative program of the Standing Committee of the 12th National People's Congress and its drafting was led by the Environmental Protection and Resources Conservation Committee of NPC.

As early as 1980s, major western industrial countries have

launched the domestic legislation on deep-sea and seabed mining attempting to uphold and consolidate their vested interests of resources in the Area. Recently, with the even elevating strategic position of the Area especially since the issuing of the *Advisory Opinion on Responsibilities and Obligations of States Sponsoring Persons and Entities with Respect to Activities in the Area* by the Seabed Disputes Chamber of the International Tribunal for the Law of the Sea the enthusiasm of major states in the international community is running high unprecedentedly. They have stated to the Authority that they will formulate and promulgate their domestic laws on activities in the Area one after another.

Formulating the law of the sea is a crucial measure for China to improve its legal system of the sea. With a unique legal statue, the Area is not only beyond the sea waters under China's jurisdiction (internal waters, territorial seas, exclusive economic zone, and continental shelf) but also beyond other countries' jurisdiction. China's existing laws on it all only apply to the sea waters under China's jurisdiction rather than the prospecting and exploration of the resources and environmental protection in the Area. Besides, the prospecting and exploration of resources in the Area is characterized by large investment, long duration, high risk, uncertain future revenue and others. The requirements of capital and technology for the enterprises involving in the activities in the Area are quite high. Thus, it is necessary to regulate and clarify the procedures, qualifications and incentive mechanism for the enterprises to enter the Area. All these urgently need to be determined through formulating ocean legislation.

"BLUE ENCLOSURE MOVEMENT"

Pioneering Investor

The Registration of China Ocean Mineral Resources Research and Development Association (COMRA) as the Pioneer Investor of the Area

In August 1990, Chinese delegation submitted application for R*egistration of the China Ocean Mineral Resources Research and Development Association as a Pioneer* with an application of mining contract covering 300,000 square kilometers enclosed. On March 5th 1991, the Preparatory Commission for the International Seabed Authority according to the provisions of UNCLOS divided the mining zone in China's application into two sub-zones of equal commercial values among which one area of 150,000 square kilometers will be given to the International Seabed Authority as a reserved area and the other area of the same range will be allocated to China as pioneer area. In August the same year, the then Secretary-General of the United Nations Javier Pérez de Cuéllar and the delegate and ambassador to the United Nations singed the certificate of registration. The registration of China Ocean Mineral Resources Research and Development Association (COMRA) as the pioneer investor of the Area marks that China's maritime undertaking has ushered new stage of rapid growth and China has become the fifth pioneer investor to be registered.

During the negotiation process of UNCLOS, many countries, including the United States, Germany, the United Kingdom and Japan and other countries, claimed that their enterprises and the international financial groups they participated in had already spent considerable amount of money on seabed exploration which could be considered as huge investment. They maintained that UNCLOS should admit their preparatory investment and assure certain privileges to them.

The *Resolution on Preparatory Investment in Pioneer Activities Relating to Polymetallic Nodules* (PIP) was signed on April 30th 1982 at the third United Nations Conference on the Law of the Sea. It aimed to solve problems of protecting investment on exploration and advancement of technology in the Area after UNCLOS had been passed but before it came into force.

According to this resolution, pioneer investors can only be applied under two circumstances. First, a pioneer investor should have spent US $30 million (relative to 1982 value) before January 1st, 1983 in pioneering activities, no less than 10 per cent of that amount, namely 3 million dollars should have been spent on the location, survey and evaluation of the pioneering area specified in the application for registration. The date for developing countries is is prolonged to January 1st, 1985. Pioneering activities refer to preparatory work that must be carried out before large-scale commercial exploitation. They include both prospecting and exploration.

Second, an applicant must have one or more signatory state as its proving states, proving that the applicant did spent the above-mentioned amount of money in pioneering activities. When the prospective pioneer investor was a State or state enterprise, then the state is the sponsoring country. If the applicant is international consortium, any state of the enterprises in the consortium signing UNCLOS could be the sponsoring state.

The *Resolution on the Preparatory Investment in Pioneer Activities* recognized four pioneer investors as State or State enterprises or their nationals or entities effectively controlled by them, namely France, India, Japan and the USSR and members of four consortia components, namely Belgium, Canada, Germany, Italy, Japan, Netherlands, the UK and the U.S. Therefore there are overall 11 countries entitled as pioneer investors according to UNCLOS. Later, India, China, the Interoceanmetal Joint Organization (Bulgaria, Cuba, Slovakia, Czech Republic, Poland and Russian Federation) and South Korea also gained the status of pioneer investors.

First-Round Deep-sea Enclosure Movement

COMRA and the International Seabed Authority signed the Contract for the Prospecting and Exploration of Polymetallic Nodules Resource in the Area in May 2001 in Beijing which defines the exclusive prospecting rights and mining concession priority of COMRA over the contract area of 75,000 square kilometers in the form of law. According to UNCLOS and relevant provisions of the International Seabed Authority, within eight years after gaining the pioneer area of 150,000 square kilometers COMRA gave up 30% of the total area of the pioneer area on March 5th 1996 and 20% on March 5th 1999 respectively through a large amount resources investigation and evaluations. It gave up the mining area of relatively poor quality with an area of 75,000 square kilometers in total and finally preserved a mining area of polymetallic nodules of relatively good quality covering an area

Pioneer Investors Applied for Polymetallic Nodules Mining Areas

Applicant (Sponsoring State)	Date of Submission	Date of Approval	Location
YUZHMORGEOLOGIYA (Russia)	1997	1997	The Pacific
Interoceanmetal Joint Organization (Bulgaria, Cuba, Slovakia, Czech Republic, Poland and Russian Federation)	1997	1997	The Pacific
South Korean Government	1997	1997	The Pacific
China Ocean Mineral Resources R&D Association (China),	1997	1997	The Pacific
Deep Ocean Resources Development Company (DORD, Japan)	1997	1997	The Pacific
IFREMER (France)	1997	1997	The Pacific
the Indian Government	1997	1997	The Indian Ocean

of 75,000 square kilometers. The contract of prospecting and exploration for polymetallic nodules in the Area signed between China and the Authority will expire in 2016.

In accordance with *The Agreement of Implementation* of Part XI of the UNCLOS, registered pioneer investor was entitled to request approval of a plan of work for exploration within 36 months of the entry into force of the UNCLOS. YUZHMORGEOLOGIYA (Russia), the Interoceanmetal Joint Organization (Bulgaria, Cuba, Slovakia, Czech Republic, Poland and Russian Federation), the South Korean government, the China Ocean Mineral Resources R&D Association (COMRA), the Deep Ocean Resources Development Company (DORD, Japan), IFREMER (France) and the Indian government all submitted their applications for prospecting on August 19th, 1997 to secretary-general of the Administration.

At the 22nd session of the Authority on August 27th 1997, the Council, acting on the recommendation of the Legal and Technical Commission, the plans of work for exploration submitted by the seven registered pioneer investors were considered to be approved and requested the Secretary-General to take the necessary steps to issue the plans of work in the form of contracts and submit the draft to each of the registered pioneer investors in August 2000.

On March 29th 2001, the Authority signed contracts with the YUZHMORGEOLOGIYA (Russia), the Interoceanmetal Joint Organization (Bulgaria, Cuba, Slovakia, Czech Republic, Poland and Russian Federation) and South Korea. The contract between the Authority and COMRA was signed in Beijing on May 22nd 2001. The contract between the Authority and DORD was signed at Kingston on June 20th 2001. On the same day, the contract between the Authority and IFREMER/AFERNOD was signed. On March 24th 2002, the Authority signed contract with the Indian government in Kingston, Jamaica.

All the contract areas are in Clarion-Clipperton Fracture Zone (CCFZ) but that of the Indian government is located in the northern part of the Indian Ocean.[1] In accordance with UNCLOS and *Regulations on the Prospecting and Exploration for Polymetallic Nodules*, each contractor obtained a 15-year prospecting contract area with 7.5 square kilometers in the Area. Contractors have exclusive right of prospecting in the allocated areas during this period. Barring special circumstances, contractors can start their commercial exploitation when their contracts expire.

Second-Round Deep-sea Enclosure Movement

People's passion for prospecting and exploration in the Area cooled down after the implementation of UNCLOS and *the Agreement of Implementation* since the future of the exploration

1. The Clarion-Clipperton Zone is located in the eastern central Pacific, to the south and south-east of the Hawaiian Islands. The geographical limits of the management area have been taken to be the area beyond national jurisdiction contained within a box approximately 0°-23°30'N x 115°W-160°W. The limits of Clarion-Clipperton Zone to the south and the north are the ENE-WNW Clarion-Clipperton Fracture. Its area is approximately 4.5 x 106 square kilometers (km^2). *Environment Management Plan for the Clarion-Clopperton Zone*, ISBA/17/LTC/7, 14.

of polymetallic nodule is not clear. Besides the pioneer investors mentioned above, Germany was the only country applying for prospecting and exploring polymetallic nodules in 2005.

With the development of deep-sea technology and the discoveries of new resources like cobalt-rich crusts and polymetallic sulphides in the Area, international attention was again paid to the Area. During the formulation process of the *Regulations on Prospecting and Exploration for Cobalt-Rich Ferromanganese Crusts in the Area* and *Regulations on Prospecting and Exploration for Polymetallic Sulphides in the Area*, some states intensified their efforts in prospecting and exploring new resources such as cobalt-rich crusts and polymetallic sulphides. After two regulations were passed by the Authority, the Secretary-General received new

SEVAN 650 II, the most advanced ultra-deepwater cylindrical offshore drilling platform in the world was successfully constructed in COSCO Shipyard (Nantong) owned by COSCO Shipyard Group and renamed as "SEVAN BRAZIL", which would be delivered and put into use very soon.

applications for mining areas. Meanwhile, the "old" resource of the Area, polymetallic nodules was once again taken seriously by some countries and commercial companies. Thus the second round enclosure movement began.

Application for Polymetallic Nodules

On July 21st 2005, the Secretary-General of the International Seabed Authority received a new application for the approval of a plan of work for exploration for polymetallic nodules in the Area. The application was submitted pursuant to the *Regulations on Prospecting and Exploration for Polymetallic Nodules in the Area* (the *Regulations*) by the German Federal Institute for Geosciences and Natural Resources on behalf of Germany. At its 11th session, the Council of the Authority approved of this application. On July 19th 2006, Germany signed the contract with the Authority. This is the first application proposed since UNCLOS came into force.

In 2008 the Republic of Nauru sponsored an application by Nauru Ocean Resources Inc. for a plan of work to explore for polymetallic nodules in the Area. The area under application covers a total combined surface of 74,830 square kilometers,[1] located within the Clarion-Clipperton Fracture Zone in the central Pacific Ocean reserved for the Authority. In 2011, the Authority approved of its application and signed a contract with NORI in the same year.

In the same year, Tonga Offshore Mining Ltd. Submitted an application, requesting for a total area of 74,713 square kilometers

1. *Nauru Ocean Resources Inc. Application for Approval of a Plan of Work for Exploration* (ISBA/14/LTC/L.2)

The Second Round Application for Polymetallic Nodules			
Applicant (Sponsoring State)	**Date of Submission**	**Date of Approval**	**Location**
German Government	2005	2005	The Pacific
Nauru Ocean Resources Inc (Nauru)	2008	2011	The Pacific
Tonga Offshore Mining Ltd. (Tonga)	2008	2011	The Pacific
UK Seabed Resources Ltd. (UK)	2012	2012	The Pacific
Marawa Research and Exploration Ltd. (Pakistan)	2012	2012	The Pacific
G-TEC Sea Mineral Resources NV (Belgium)	2012	2012	The Pacific
UK Seabed Resources Ltd. (UK)	2013	2014	The Pacific
Ocean Mineral Singapore Pte Ltd (Singapore)	2013	2014	The Pacific
the Cook Islands Investment Corporation (The Cook Islands)	2013	2014	The Pacific

to explore polymetallic nodules in the Clarion Clipperton Fracture Zone in the central Pacific reserved for the Authority under the sponsorship of Kingdom of Tonga. The Authority approved of this application and signed the contract in 2012.

On May 23rd 2012, the Secretary-General of the International Seabed Authority received an application for the approval of a plan of work for exploration for polymetallic nodules in the Area. The application was submitted by UK Seabed Resources Ltd. sponsored by the United Kingdom. The application covers a total surface area of approximately 116,000 square kilometers located in the eastern part of the Clarion-Clipperton Fracture Zone in the Pacific Ocean. The Authority approved of the application in 2012 and signed the contract in February, 2013.

On May 30th 2012, the Authority received an application for the approval of a plan of work for exploration for polymetallic nodules in the Area. The application was submitted by Marawa Research and Exploration Ltd. sponsored by Pakistan. The area under application covers a total surface area of 74,990 square kilometers and is located within the areas reserved for the Authority. In the same year, the Authority approved its application.

On May 31st, 2012, the Authority received an application for the approval of a plan of work for exploration for polymetallic nodules in the Area. The application was submitted by G-TEC Sea Mineral Resources NV sponsored by Belgium. The application covers a total surface area of 148,665 square kilometers and is located in the central part of the East Clarion-Clipperton Fracture Zone in the Pacific Ocean. In January 2013, the Authority signed the contract with G-TEC.

In February 2013, the Authority received and approved of

applications for the approval of plans of work for exploration for polymetallic nodules in the Area from resourced limited companies of the UK, Singapore and the Cook Island.

On August 8th 2014, China Minmetals Corporation submitted to the International Seabed Authority the exploration application of polymetallic nodules with Chinese government as the sponsor. The application area is located in the reserved area in the Clarion-Clipperton Fracture Zone (CCFZ) in central Pacific. The Legal and Technical Commission of the Authority will review on this application at the meeting in February 2015.

Established in 1950, China Minmetals Corporation is a large-scale diversified metals and mining company engaged in the development, production and trading of metals and minerals, and comprehensive services as well as other businesses. With regard to the exploitation of marine mineral resources, the corporation possesses advantaged research basis and development capacity. It has established a state key laboratory of the development of utilization of deep-ea mineral resources and boasted of professional technological capacity of the national advanced level in the field of marine mineral resources development. Currently, the corporation has already conducted key research in seabed prospecting and exploration, transporting and dressing and smelting processing test and other links. Its subgroup Changsha Mining and Smelting Research Institute is also conducting research on deep-sea mining of resources such as polymetallic nodules, polymetallic sulphides, and cobalt-rich crusts. The participation of large-scale state-owned enterprises like China Minmetals Corporation in the prospecting and exploration of the resources in the International Seabed Area plays a vitally significant role in enhancing China's international competitiveness in relevant fields.

Application for Polymetallic Sulphides

The China Ocean Mineral Resources Research and Development Association (COMRA) submitted the first international application to explore a polymetallic sulphide ore deposit located in Southwester Indian Ocean Ridge to the International Seabed Authority (ISA) in May 2010. The ISA Council approved the application in 2011 and with COMRA they signed the contract in Beijing.

In December 2010, the Ministry of Natural Resources and the Environment of the Russian Federation submitted the second application to explore polymetallic sulphides in the Mid-Atlantic

Application for Polymetallic Sulphides

Applicant (Sponsoring State)	Date of Submission	Date of Approval	Location
COMRA (China)	2010	2011	the Indian Ocean
the Russian Government	2010	2011	the Atlantic Ocean
Korean Government	2012	2012	the Indian Ocean
IFREMER (France)	2012	2012	the Atlantic Ocean
the Indian Government	2013	2014	the Indian Ocean
German Government	2013	2014	the Indian Ocean

Ridge. The Authority approved of the application in 2011 and signed the contract in October, 2012.

In May 2012, Ministry of Land Transport and Maritime Affairs of South Korea submitted an application for a plan to explore the polymetallic sulphides in the centre of the Indian Ocean on behalf of the Korean government. The Authority approved of this application in the same year.

In June 2012, IFREMER submitted an application for the approval of a plan of work for exploration for polymetallic sulphides along the Volcano Ridge in central Atlantic. In the same year, the Authority approved of this application.

In March 2013, Indian government submitted its application to prospect polymetallic sulphides in the centre of the Indian Ocean. The area it applied for covered a total surface area of 10,000 square kilometers and the Authority approved of such application in July 2014.

In December 2013, the German government applied for a plan of work for prospecting for polymetallic sulphides in the centre of the Indian Ocean and it was approved of by the Authority in July 2014.

Application for Cobalt-rich Crusts

In July 2012, COMRA submitted the first application for the approval of a plan of work for exploration for cobalt-rich ferromanganese crusts in the Area. The place it applied for is located in the seamounts of the West Pacific which covers an area of 3,000 square kilometers. The Authority approved its application in 2013.

Application for Cobalt Rich Ferromanganese Crusts

Applicant (Sponsoring State)	Date of Submission	Date of Approval	Location
COMRA (China)	2012	2013	The Pacific
Japan Oil, Gas and Metals National Corporation (Japan)	2012	2013	The Pacific
Russian Government	2013	2014	The Pacific
CPRM (Brazil)	2013	2014	The Atlantic

In August 2012, the Japanese government sponsored Japan Oil, Gas and Metals National Corporation to apply for the approval of a plan of work for exploration for cobalt-rich ferromanganese crusts located in the seamounts of the West Pacific. The Authority approved the application in 2013.

In February 2013, the Russian Federation submitted to the Secretary-General of the International Seabed Authority an application for approval of a plan of work for exploration for cobalt-rich ferromanganese crusts in the Area. The area it applied for is located in Magellan Mountain in the Pacific Ocean, which covers a total surface of 6,000 square kilometers. The Authority approved of its application in July, 2014.

In December 2013, the Authority received an application for approval of a plan of work for exploration for cobalt-rich ferromanganese crusts in the Area submitted by Companhia de

Pesquisa de Recursos Minerais (CPRM), which is a Brazilian State Enterprise. The application was sponsored by the Government of the Federative Republic of Brazil. The application area is in the South Atlantic and the Authority approved of the application in July 2014.

Conclusion: A Prospect for China's Maritime Undertaking

To meet the need of developing China's maritime undertaking, under the support of related councils and units, China Ocean Mineral Resources R&D Association has improved the network platform and established China Ocean Sample Repository, Ocean Biological Gene Research Center, Marine Earth Science and Environmental Research Center and Marine Exploration Technology and Deep-Sea Science Research Base one after another. These platforms have provided essential support for the development of China's maritime undertaking. In January 2007, the State Council approved of the establishment of a national deep-sea base. This makes China the fifth country to have technology supported base after Russia, the United States, France and Japan,

a Schematic Map of National Deep Sea Base

marking the construction of a national multi-functional and fully open public service platform. This is paramount for China to improve its ocean investigation level.

Through a review on the negotiation process of regulations of the Area and legal regime of the Area in UNCLOS, it is evident that the Area is indeed a special political and geographical unit. Resources in the Area are "common heritage of mankind". All States have the rights to exploit and utilize resources in the Area, as well as the responsibility and obligation to protect the resources and the Area itself. The limits of the Area are closely connected to the outer limit of continental shelf beyond 200 nautical miles. The Area will get smaller if the outer continental shelf extends its limit. Coastal countries shall avoid promoting irrational or even illegal proposals so as not to ruin the common heritage of mankind of the Area. If any state proposes such claim, the Authority that runs the common heritage of mankind on the behalf of all human beings should take actions immediately in order to protect the interests of the international society as a whole.

The principle of "common heritage of mankind"as the guiding principle for prospecting and exploration is rooted in its special history and time background. Back to 1960s to 1970s, developing countries began to express their voice on the international stage. Strongly opposed to hegemonism, they pursued democratization of international relations and proposed to establish a new international economic order. Under this circumstance, during the negotiation process of the legal regime related with the Area developing countries strongly opposed to some claims of developed countries such as the Area is and all states have the freedom to prospect and explore the resources in the Area and so on. The principle of

Xisha Island in Hainan Province

"common heritage of mankind" principle was a product in that background.

From UNCLOS to the *Agreement of Implementation*, and from the *Regulations on Prospecting and Exploration for Polymetallic Nodules* and *Regulations on Prospecting and Exploration for Sulphides* to *Regulations on Prospecting and Exploration for Cobalt-rich Crusts*, with nearly 30 years' development the legal regime of the Area has developed into maturity. The passion of the international community for exploring and exploiting resources in the Area is regained. A new round of "blue enclosure movement" will awaken the world. Since the first group of contracts approved of by the Authority will be expired in 2016, the Authority is now actively formulating new regulations for exploitation to setting rules and regulations for commercial exploitation of the resources in the Area. It is foreseeable that commercial exploitation of resources in the Area will definitely be realized in the near future.

China attained great achievements in the two rounds of "blue enclosure movements" and obtained three contract areas for exploration and exploitation thanks to the national attention attached to it and the efficient organization of departments concerned. This has laid a sound foundation for further activities in the Area. Maritime undertaking is a systematic cause. Aiming to occupy an advantage in the increasingly competitive activities in the Area, China must efficiently utilize the resources in the Area in order to gain real interests. Therefore, there are still lot to be improved with regard to resource exploration depth, basic capacity building, R&D of deep-sea equipment and technology and so on. China embarks upon an arduous journey of maritime undertaking.